Synthesis of New [2.2]Paracyclophane Derivatives for Application in Material Sciences

Zur Erlangung des akademischen Grades eines/einer

DOKTORS/DOKTORIN DER NATURWISSENSCHAFTEN

(Dr. rer. nat.)

von der KIT-Fakultät für Chemie und Biowissenschaften

des Karlsruher Instituts für Technologie (KIT)

genemigte

DISSERTATION

von

M. Sc. Eduard Spuling

aus Pawlodar

Dekan: Prof. Dr. Reinhard Fischer

Referent: Prof. Dr. Stefan Bräse

Koreferent: Prof. Dr. Uli Lemmer

Tag der mündlichen Prüfung: 07. Mai 2019

Band 82
Beiträge zur organischen Synthese
Hrsg.: Stefan Bräse

Prof. Dr. Stefan Bräse
Institut für Organische Chemie
Karlsruher Institut für Technologie (KIT)
Fritz-Haber-Weg 6
D-76131 Karlsruhe

Bibliographic information published by the Deutsche Nationalbibliothek

The Deutsche Nationalbibliothek lists this publication in the Deutsche Nationalbibliografie; detailed bibliographic data are available in the Internet at http://dnb.d-nb.de

ISBN 978-3-8325-4997-8
ISSN 1862-5681

Logos Verlag Berlin GmbH
Comeniushof, Gubener Str. 47,
10243 Berlin
Tel.: +49 030 42 85 10 90
Fax: +49 030 42 85 10 92
INTERNET: http://www.logos-verlag.de

"Scientific research is one of the most exciting and rewarding of occupations.

It is like a voyage of discovery into unknown lands, seeking not for new territory but for new

knowledge. It should appeal to those with a good sense of adventure."

–   Frederick Sanger

Honesty Declaration

This work was carried out from January 1st 2016 through March 27th 2019 at the Institute of Organic Chemistry, Faculty of Chemistry and Biosciences at the Karlsruhe Institute of Technology (KIT) under the supervision of Prof. Dr. Stefan Bräse.

Die vorliegende Arbeit wurde im Zeitraum vom 1. Januar 2016 bis 27. März 2019 am Institut für Organische Chemie (IOC) der Fakultät für Chemie und Biowissenschaften am Karlsruher Institut für Technologie (KIT) unter der Leitung von Prof. Dr. Stefan Bräse angefertigt.

Hiermit versichere ich, EDUARD SPULING, die vorliegende Arbeit selbstständig verfasst und keine anderen als die angegebenen Hilfsmittel verwendet sowie Zitate kenntlich gemacht zu haben. Die Dissertation wurde bisher an keiner anderen Hochschule oder Universität eingereicht.

Hereby I, EDUARD SPULING, declare that I completed the work independently, without any improper help and that all material published by others is cited properly. This thesis has not been submitted to any other university before.

German Title of this Thesis

Darstellung neuer [2.2]Paracyclophanderivate für materialwisschenschaftliche Anwendungen

Table of Contents

Kurzzusammenfassung

Seit der Entdeckung des [2.2]Paracyclophans vor exakt 70 Jahren durch Brown und Fathing[1] entfachte diese Verbindung großes Interesse in der Wissenschaft. Die zwei zueinander coplanar angeordneten Benzole mit einem sehr geringen Abstand von 3.09 Å,[2-3] die durch zwei zueinander in *para*-Position stehenden Ethylbrücken verbrückt werden, ergeben ein gespanntes und verzerrtes "*bent and battered*"[2] Grundgerüst, so wie es in der klassischen, nichtcyclophanischen Aromatenchemie unbekannt ist. Die besonderen Eigenschaften dieses Moleküls sind I) die einzigartige transannulare elektronische Kommunikation der zwei π-Systeme ermöglicht durch den besonders geringen Benzolringabstand, [4-5] II) die Präsenz einer planaren Chiralität bei Funktionalisierung und III) der beträchtliche sterische Raumanspruch durch die Ethylbrücken und den coplanar liegenden zweiten Benzolring. Aufgrund dieser besonderen Eigenschaften ist dieses Molekül synthetisch anspruchsvoll, da die verbogene und gespannte Struktur die Chemie in besonderer Weise beeinflusst. Bis zum heutigen Tage wurde das [2.2]Paracyclophangerüst hauptsächlich in der Katalyse als chiraler Ligand erforscht. Im Gegensatz dazu sind Untersuchungen zu materialwissenschaftlichen Anwendungen sehr viel seltener berichtet.[6]

Daher ist das Ziel dieser Arbeit synthetische Zugänge zu mehreren [2.2]Paracyclophan-basierten Derivaten zu finden die es mehreren Materialwissenschaftlichen Anwendungen und der Supramolekularen Chemie ermöglicht die transannulare Kommunikation, den besonderen sterischen Raumanspruch und die Chiralität auszunutzen.

Die transannulare elektronische Kommunikation wurde für das Design von neuartigen Luminophoren genutzt um sogenannte *thermally activated delayed fluorescence* (TADF) Emitter darzustellen. Es konnten aus einer Reihe von zehn dargestellten Zielstrukturen zwei blau emittierende Moleküle gewonnen werden die den TADF Mechanismus zur effizienten Lichterzeugung nutzen.

Der besondere Raumanspruch wurde genutzt um [2.2]Paracyclophan-basierte (1,4)Carbazolophan-analoga zum im Emitterdesign gängigen Carbazol darzustellen. Der dargestellte TADF Emitter besaß die erwartete erhöhte Torsion, welche TADF Emission ermöglichte. Dies erlaubte eine OLED mit einem EQE$_{max}$ von 17% herzustellen.

Desweiteren wurde die Untersuchung des Chiralitätstransfers von einem Molekül mittels chemischer Gasphasenabscheidung generierten Poylmerfasern durch die geeigneter chiraler [2.2]Paracyclophan-basierter Monomere ermöglicht.

Schließlich konnte durch die Synthese eines geeigneten [2.2]Paracyclophanderivates die Voraussage, dass das sphärische Grundgerüst ein vielversprechender Gast in der supramolekularen Wirts-Gast Chemie ist.

Während all dieser synthetischen Problemstellungen wurden die Besonderheiten der [2.2]Paracyclophan weiter erforscht. So wurde eine neuartige *para*-selektive C–H Aktivierung und die Darstellung eines neuartigen (1,7)Carbazolophans untersucht.

Abstract

Since the first discovery of [2.2]paracyclophane by Brown and Farthing exactly 70 years ago[1] this molecule has sparked tremendous interest in the scientific community. The two coplanar phenyl rings are in close proximity of 3.09 Å,[2-3] fixed by two short ethyl bridges in *para*-position resulting in a "*bent and battered*"[2] molecular skeleton which is unknown in conventional non-cyclophane aromatic chemistry. The key features of this molecule are I) the unique electronic situation enabling transannular electronic communication of the two π -systems,[4-5] II) the presence of planar chirality upon substitution and III) the considerable spherical bulk induced by the adjacent ethyl bridge and second co-planar phenyl ring. Despite these unique features, the bent and therefore unfavored conformation influences the chemistry of this molecule dramatically, making the synthetic derivatization a challenging endeavor. Up until now the [2.2]paracyclophane scaffold has predominantly been investigated in the context of catalysis as a chiral ligand,[7-9] utilizing the chirality and steric hindrance of this molecule. In contrast, investigations regarding material sciences are far less explored.[6]

Therefore, the aim of this thesis was to establish a synthetic access to a variety of [2.2]paracyclophane-based derivatives for different applications in material sciences and supramolecular chemistry exploiting transannular communication, steric demand and chirality.

For the design of luminescent molecules, the through-space electronic communication of the [2.2]paracyclophane scaffold was taken advantage of, and the promising thermally activated delayed fluorescence (TADF) mechanism for efficient light generation was established, giving two blue emitters out of ten synthesized target structures for which the TADF relaxation principle was verified.

The exceptional steric demand was utilized to generate a [2.2]paracyclophane-based (1,4)carbazolophane analogue of the common carbazole group for TADF emitter design. The additional twist proved to be a significant factor in the turn-on of the TADF relaxation mechanism, ultimately yielding an OLED with an EQE_{max} of 17%.

Furthermore, for the chemical vapor deposition (CVD) process, the influence of molecular chirality on the helicity of polymerized fibers was evaluated. For this suitable chiral [2.2]paracyclophane derivatives were provided and could be shown that chirality can be transferred from the molecular level to a micrometer environment.

Ultimately, the spherical skeleton proved to be a promising guest molecule in supramolecular chemistry when functionalized properly.

During all these synthetic challenges, the peculiarities of this intriguing structure were observed and investigated, leading to the discovery of a unprecedented *para*-selective C–H activation and a synthetic access to the novel (1,7)carbazolophane.

1 Introduction

"Chemical synthesis is uniquely positioned at the heart of chemistry, the central science, and its impact on our lives and society is all pervasive."[10] was written by E. J. Corey in his Nobel Lecture in 1991 in order to emphasize the role synthesis has. And G. S. Hammond said that *"The most fundamental and lasting objective of synthesis is not production of new compounds, but production of properties."*[11] To comply with both valid statements, it is fundamental for present-day organic synthesis to selectively plan and carry out projects in order to discover novel properties in both synthesis and material sciences. In this regard, this thesis focused exclusively on the chemistry and application of a structurally exceptional scaffold, namely the [2.2]paracyclophane (**1**).

This compound is one of the most noteworthy and intriguing representatives of the structural class of cyclophanes (Figure 1) for which systematic studies were pioneered by Cram[2] and Hopf.[12]

$$\left(\begin{matrix} (CH_2)_n \\ \\ Arene \end{matrix}\right) \qquad \left(\begin{matrix} Arene_a \\ (CH_2)_n \qquad (CH_2)_m \\ Arene_b \end{matrix}\right)$$

Figure 1. Conceptional representation of the structural class of cyclophanes.

This class consists of a **cyclic** arrangement of at least one aromatic (**phenyl**) and at least one aliphatic bridge (**alkane**) and can therefore include pure hydrocarbon compounds while aliphatic or aromatic heteroatoms are common. In Figure 2, a small selection of remarkable representatives of this structural class is given. Apart from the later discussed [2.2]paracyclophane (**1**), this structural class is ubiquitous in chemistry, e.g. "Stoddart's blue box" (**2**),[13] which is frequently used in supramolecular chemistry to prepare catenanes,[14] or the natural product myricanol(**3**),[15] which can be isolated from *Myrica cerifera* and is used as an anesthetic in folk medicine. Even the Vogel aromatic **4**, a bridged [10]annulene, can be considered a cyclophane.[16]

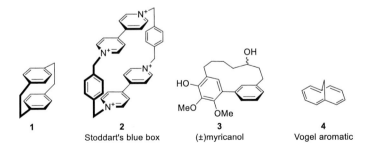

1	**2**	**3**	**4**
	Stoddart's blue box	(±)myricanol	Vogel aromatic

Figure 2. Selected examples of cyclophanes.

In contrast to non-cyclic compounds, the often short bridges in cyclophanes are able to impede or completely prohibit free rotation and therefore can constrain these molecules into unusual and thermodynamically disfavored conformations.[17] The biphenyl bond in compound **3**, for example, is bent out of the planes of both phenyl rings.[15] The investigation of structural peculiarities, and the special chemical properties that derive, represent an exciting field of synthetic chemistry, which also includes propellanes.[18] The archetypical representative of strained cyclophanes is the [2.2]paracyclophane (**1**) which was perfectly paraphrased by Cram as "*bent and battered benzene rings*".[2]

1.1   The [2.2]Paracyclophane

1.1.1   Structure of [2.2]Paracyclophane

First reported by Brown and Farthing,[1] the [2.2]paracyclophane **1** is a strained system with the benzene rings (also called "decks") significantly bent out of the expected ring plane (Figure 3). This is caused by the short ethyl bridges, which force the bridgehead atoms of the benzene rings out of the expected aromatic planarity at an angle of 13°. Additionally, the two stacked rings are declipsed with equilibration observed even at 93 K.[3] The forced boat-shaped structure results in a reduced overlap of the p-orbitals and thus leads to a diminished aromaticity. Furthermore, the distance between the two benzene rings amounts to 3.09 Å, which is 92% of the layer distance reported for graphite (3.35 Å).[19] Therefore, it is commonly accepted that there is interaction between the two π-systems of both benzene moieties. These special properties can be seen in ^{1}H NMR, where the aromatic protons are shifted upfield to $\delta =$ 6.50 ppm, while in benzene they are observed at $\delta = 7.26$ ppm.[2-3] The central C–C single bond in the ethylene bridge is 6% longer than "normal" single bonds.[20] The ring strain amounts to 134 kJ/mol, and rotation of the benzene units is not possible due to the short bridges. This conformational stability allows the synthesis and separation of enantiopure derivatives.[21] All these structural abnormalities correlate with peculiar chemical properties.

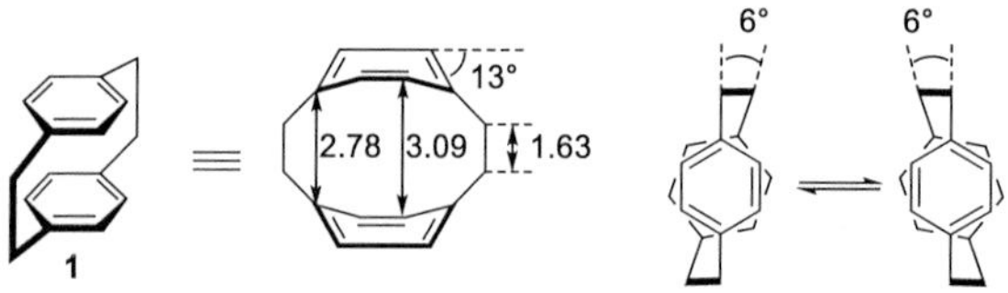

Figure 3. Structural parameters of [2.2]paracyclophane. Distances are given in Ångström.[22]

Since naming of cyclic compounds is rather difficult, the official IUPAC nomenclature for cyclophanes is inconvenient. Therefore, Vögtle *et al.* developed a specific cyclophane nomenclature, which is based on a core-substituent ranking.[23] The core structure is named according to the length of the aliphatic bridges in square brackets (e.g. [n.m]) and the benzene substitution patterns (*ortho, meta* or *para*).

[2.2]Paracyclophane belongs to the D_{2h} point group, which is broken by the first substituent, resulting in two planar chiral enantiomers. By definition, the arene bearing the substituent is set to a chirality plane (Figure 4, top), and the first atom of the cyclophane structure outside the plane and closest to the chirality center is defined as the "*pilot atom*". If both arenes are substituted, then the substituent with higher priority according to the Cahn-Ingold-Prelog (CIP) nomenclature is preferred.[24] The stereodescriptor is determined by the sense of rotation viewed from that pilot atom and to indicate the planarity of the chiral center, a subscripted *p* is added.[25]. To correctly describe the positions of the substituents, an unambiguous numeration is needed, but unfortunately, the numbering of the second arene is not consistent in literature. The numbering of the arenes should follow the sense of rotation determined by CIP, but another description based on the benzene substitution patterns is preferred for disubstituted [2.2]paracyclophanes, (Figure 4, bottom) with the descriptor "pseudo" when the substituent is at the second arene.

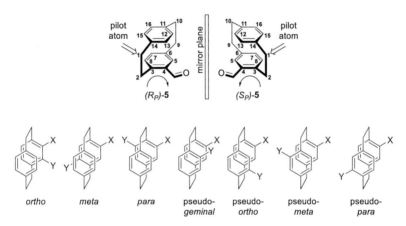

Figure 4. Top: Nomenclature and chirality descriptors exemplified by 4-formyl[2.2]paracyclophane (**5**). Bottom: Available aromatic substitution pattern of disubstituted [2.2]paracyclophane with relative nomenclature.

1.1.2   Synthesis of [2.2]Paracyclophane

The first synthesis of [2.2]paracyclophane was carried out in 1948 by Szwarc during pyrolytic studies of C–H bond energies of xylenes without further investigation of the products formed,[26] which was done one year later by Brown and Farthing.[1] Nevertheless Szwarc proposed that the comparatively fast decomposition of *p*-xylene (**6**) yields a quinodimethane diradical species **7** (Scheme 1), that can either dimerize to give [2.2]paracyclophane (**1**) or form polymers. As indicated, this process is in principle reversible.

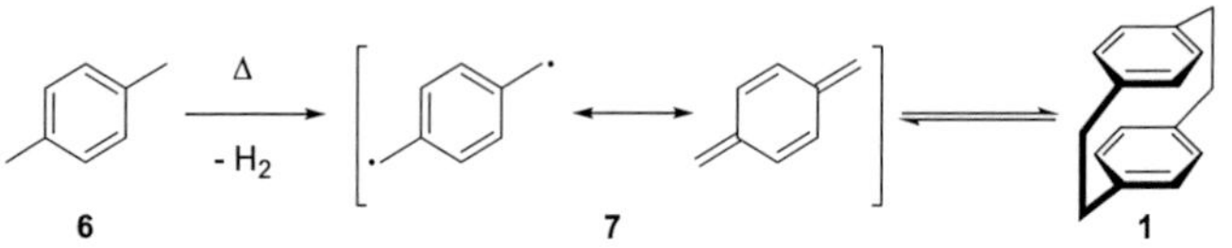

Scheme 1. Pyrolysis of *p*-xylene (**6**) to the quinodimetane (**7**) intermediate, which dimerizes to yield [2.2]paracyclophane (**1**).

In 1951, Cram *et al.* began thorough research on the aromatic anomalies, which strained macrocyclic compounds exhibit. Therefore, a general method was developed to access [n.m]paracyclophanes by intramolecular Wurtz coupling (Scheme 2).[27]

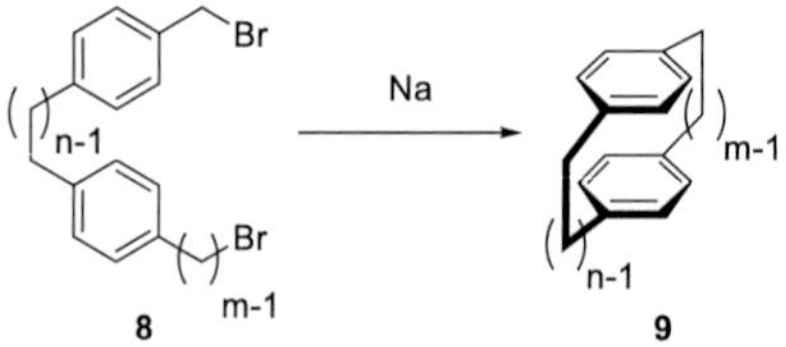

Scheme 2. First targeted synthetic access to [n.m]paracyclophanes **9** including the [2.2]-representative **1** by Cram *et al.*[27]

Another route uses the quinodimethane intermediate **7** generated *via* 1,6-Hofmann elimination (Scheme 3) of suitable precursors such as **10** which undergoes dimerization under less harsh conditions to yield the desired [2.2]paracyclophane (**1**).[28-29]

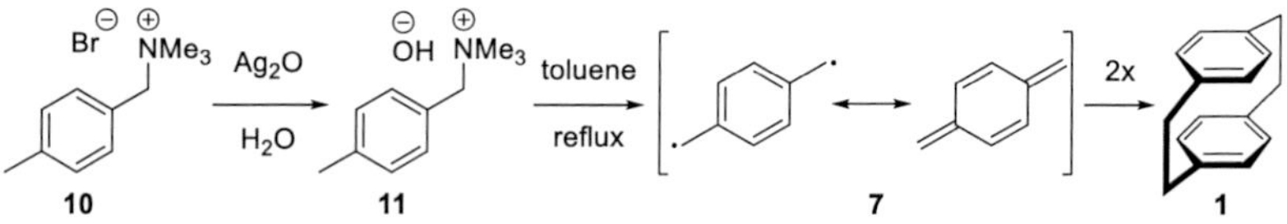

Scheme 3. 1,6-Hofmann elimination generates quinodimethane intermediate **7** to access [2.2]paracyclophane (**1**).[28]

A further approach employs the [3.3]dithiacyclophane **14** which can be easily prepared from the dithiol **12** and the dibromide **13**. This larger cyclophane undergoes ring contraction *via* sulfur extrusion through Pummerer,[30] or Stevens rearrangement.[31] This particularly elegant method is extraordinarily suitable to access [2.2]paracyclophanes with functionalizations at the aliphatic ethyl bridges, which are key intermediates to [2.2]paracyclophane-1,9-diene,[30] or to access disparately substituted phane decks, such as the [2.2]pyridinophane, which exhibits one benzene and one pyridine unit.[32]

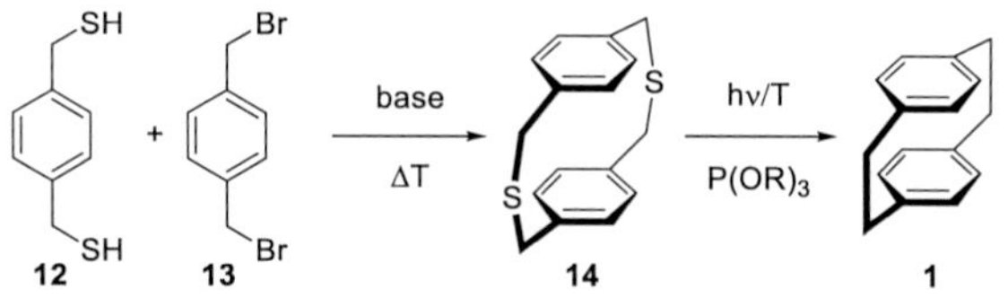

Scheme 4. Access to [2.2]paracyclophane (**1**) *via* sulfur extrusion of [3.3]dithiacyclophane **14**.

1.1.3 Reactivity of [2.2]Paracyclophane

1.1.3.1 Initial Functionalization

As an aromatic compound, [2.2]paracyclophane (**1**) is readily functionalized by electrophilic aromatic substitution reactions (Scheme 5), such as nitration, acylation, formylation and bromination, which give access to initial functionalized derivatives in good yields.

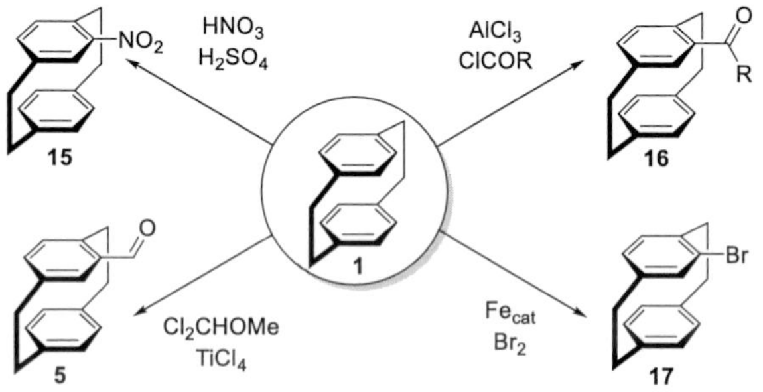

Scheme 5. Initial functionalization strategies in [2.2]paracyclophane chemistry.[33]

Due to the close vicinity of the benzene rings, a transannular directing effect for a second functionalization *via* electrophilic aromatic substitution (S_NAr) is observed with structural carbonyl groups such as ketones, esters, amides, oxazolines, phosphine oxides, sulfones, sulfoxides and nitro groups. The substitution occurs at the pseudo-*geminal* position with nearly exclusive selectivity and can be explained by evaluation of the mechanism (Scheme 6). During S_NAr, the π-complex **18** is formed a-top the electron-rich benzene ring. In case of the mentioned functional groups, this is always the non-functionalized ring. This $η^6$-complex selectively collapses to the σ-complex **19** (Wheland intermediate), because the *ipso*-hydrogen stands pseudo-*geminal* to the carbonyl function and is therefore readily deprotonated to give the intermediate **20**.

1.1.3.2   Strategies Towards Difunctionalized Derivatives

16   **18**   **19**   **20**   **21**

π-complex   σ-complex

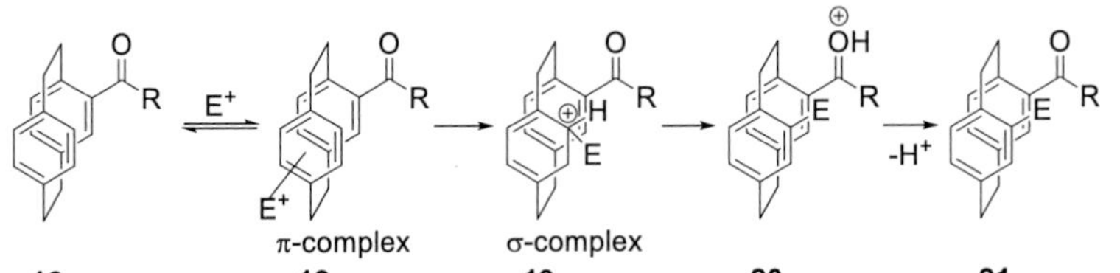

Scheme 6. Transannular directive effect of carbonyl structural groups to pseudo-*geminal* selective second functionalization *via* electrophilic aromatic substitution.[33]

Furthermore, bromination can be performed numerous times to yield dibromo[2.2]paracyclophanes, but does yield a mixture of different-pseudo-dibrominated isomers, which require tedious separation (Scheme 7).[34] The centrosymmetric pseudo-*para* dibromide **22** can be easily separated from the remaining mixture of isomers due to its low solubility and is therefore a well-used intermediate to difcunctionalized derivatives. Optimized reaction conditions lead to a 34% yield of **22**.[33]

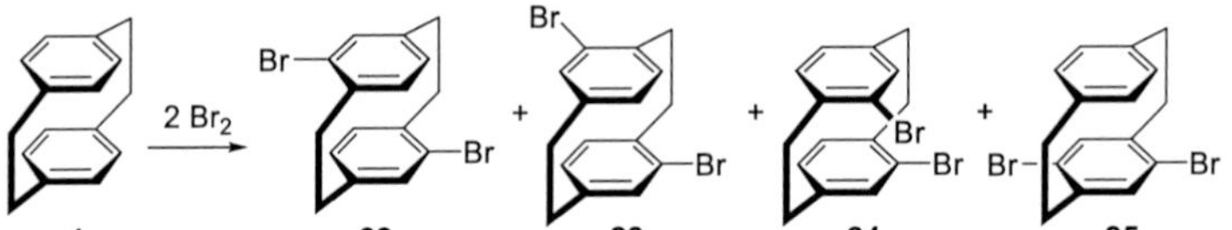

1   **22**   **23**   **24**   **25**
           pseudo-*para*   pseudo-*meta*   pseudo-*ortho*   *para*
              26%   6%   16%   5%

Scheme 7. Dibromination of [2.2]paracyclophane with isolated yields reported by Cram *et al.*[34]

In contrast to the iron catalyzed or non-catalyzed access to aromatic bromide **17** or **22–25**, functionalization at the aliphatic ethyl bridges can be achieved by photochemically initiated bromination (Scheme 8). De Meijere *et al.* reported tetrabromination to yield a mixture of 1,1,9,9- and 1,1,10,10-tetrabromides **26** and **27**.[35]

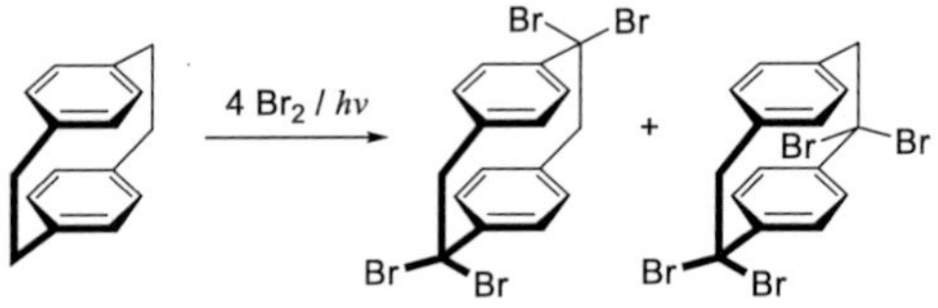

1   **26**   **27**

Scheme 8. Aliphatic bromination under photochemical conditions.[35]

The afore mentioned synthetic access *via* dimerization of quinodimethane intermediate **7** is reversible as ring strain is released, but this process occurs *via* a two-step mechanism where both tethers of the

[2.2]paracyclophane are opened homolytically in a sequential manner at high temperatures. By carefully tuning the reaction temperature, this mechanism can be set to hold at one broken ethyl bridge yielding a non-cyclic and therefore non-strained diradical, which is able to recyclize. This property is commonly exploited for the synthesis of pseudo-*ortho* dibromide **24** from the readily available pseudo-*para* dibromide **22** (Scheme 9). Despite this potential decomposition pathway of [2.2]paracyclophanes, the thermal isomerization proceeds only at significantly elevated temperatures of 200 °C.

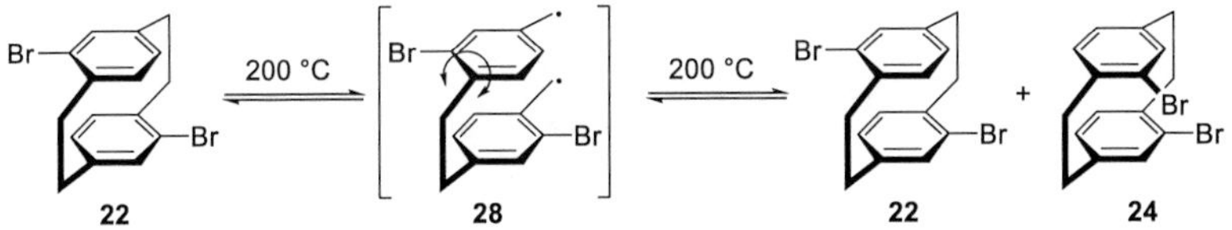

Scheme 9. Thermic isomerization of the [2.2]paracyclophane backbone leads to differently substituted stereoisomers.[33]

Further access to *ortho*-functionalized derivatives can be obtained by templated lithiation. As an example from Pelter *et al.* (Scheme 10),[36] the amide **30** can lead to both the pseudo-*geminal* bromide **31** *via* the previously discussed transannular directing effect or to the *ortho*-bromide **29** by templated lithiation.

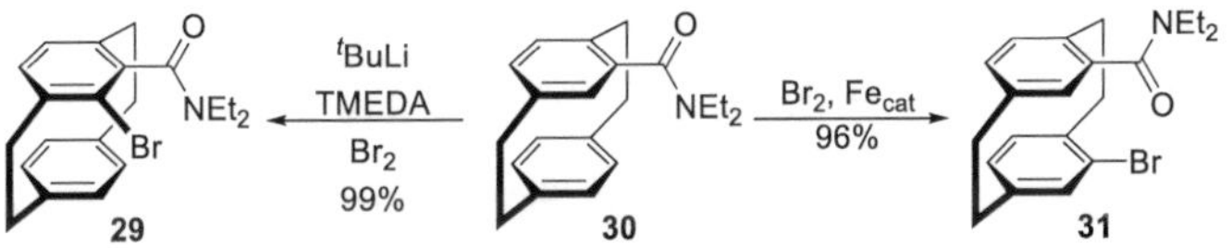

Scheme 10. Choice of regioselectivity for amide **30** can be set by selected reaction protocols (electrophilic substitution *versus* lithiation).[36]

With regard to synthetic strategies to difunctionalized [2.2]paracyclophane derivatives, pseudo-*geminal* and pseudo-*para* substitution patterns can be readily accessed by pseudo-*geminal* directive effects, and by easy separation of the dibromide **22**, respectively. The pseudo-*ortho* substitution pattern can be obtained using the dibromide **24** and *ortho*-functionalization is attained by directed lithiation.

In contrast pseudo-*meta*, *meta* and *para* substitution patterns are much more challenging to obtain, and are often accessed *via* de-novo syntheses of the whole [2.2]paracyclophane skeleton.

1.1.4   C–H Activation

C–H activation on the [2.2]paracyclophane structure was first explored by Bolm *et al.* starting in 2012. It was shown that *ortho*-selective acetoxylation mediated by Pd(OAc)₂/PIDA can be effectively directed by aldoxime ethers, ketoxime ethers and esters, 2'-pyridyl and pyrazole directing groups (Scheme 11) in good yields and selectivity under acidic conditions.[37]

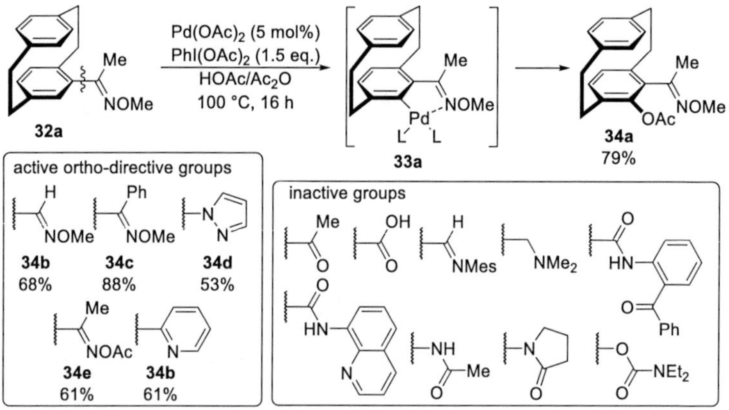

Scheme 11. Overview of performance of directive groups in *ortho*-C–H activation from Bolm *et al.*[37]

Although oximes proved to be excellent *ortho*-directing groups for Pd-catalyzed *ortho*-C–H activation, numerous other common directing groups were inactive, either because they were incompatible under the reaction conditions or too bulky.

Related to these findings, a Pd-catalyzed *ortho*-bromination/iodination procedure was described presenting a swifter access to *ortho*-functionalized intermediates although 20 mol% catalyst loading was needed to obtain 63% of products **35** (Scheme 12).[38]

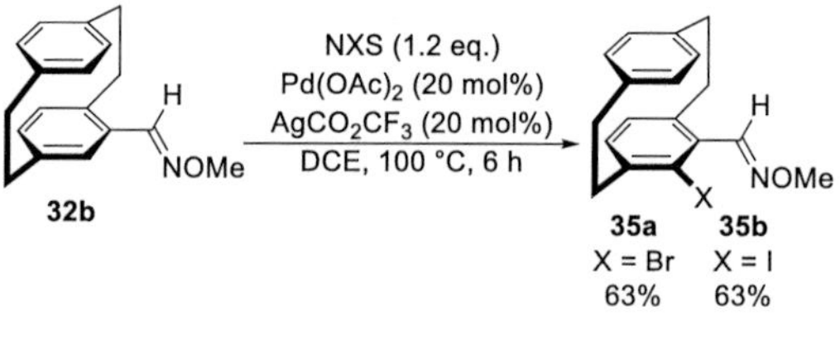

Scheme 12. Pd-catalyzed *ortho*-bromination. [38]

Bolm *et al.* further used their procedure of *ortho*-C–H activation under acidic conditions to synthesize (1,4)-bridged carbazolophanes **37** (Scheme 13) *via* oxidative cyclization under aerated conditions starting from *N*-phenylamino[2.2]paracyclophane **36**.[39] In contrast to the *ortho*-acetoxylation, a significantly increased catalyst loading of 20 mol% is needed to obtain **37** in 62% yield. Interestingly, the oxidative cyclization does not occur with a *N*-methylated derivative of **36**. Although the reaction conditions are rather harsh, this protocol is reported for a wide range of electron-rich and -poor aniline derivatives. This gives access to a very interesting class of planar chiral (1,4)-bridged carbazolophanes which are suitable for material sciences as they both exhibit an increased steric demand and planar chirality.

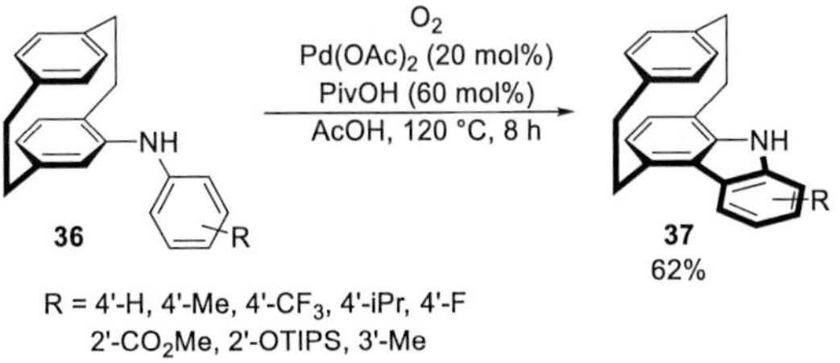

R = 4'-H, 4'-Me, 4'-CF$_3$, 4'-iPr, 4'-F
2'-CO$_2$Me, 2'-OTIPS, 3'-Me

Scheme 13. Oxidative cyclization in *ortho*-position to planar chiral (1,4)-bridged carbazolophanes.[39]

1.1.5 *para*–Activation

Bolm *et al.* also reported a *para*-functionalization of phenylamino, acetamido (and derivatives)-substituted [2.2]paracyclophanes (e.g. **38**). To obtain the product **39**, phenyliodide diacetate (PIDA) was used as the oxidant to mediate the addition of various nucleophiles (Scheme 14). Nucleophiles such as acetate, formate, methanolate, ethanolate and bromide can be successfully attached at the *para*-position in moderate to good yields. An insight into the mechanism was given, when an excess of PIDA was added, which led to the mixed benzoquinone **40** in 54% yield.[40]

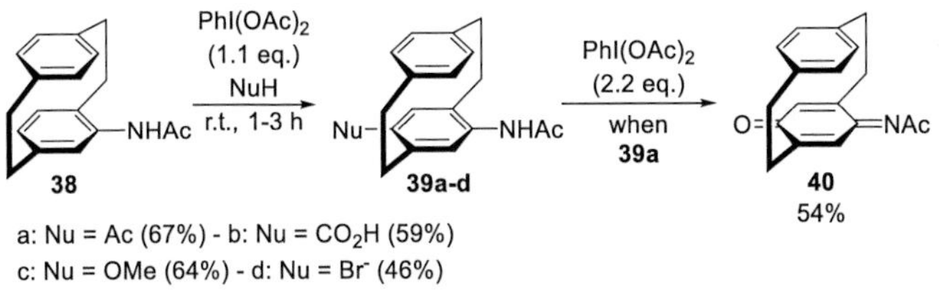

a: Nu = Ac (67%) - b: Nu = CO$_2$H (59%)
c: Nu = OMe (64%) - d: Nu = Br⁻ (46%)

Scheme 14. PIDA mediated *para*-functionalization and formation of side product **40** when an excess of oxidant is used.

In a related manner, benzoquinone **43** was reported by Cram already in 1966. It is a convenient intermediate,[41-42] since both, the precursor **41** is available, and the benzoquinone **43** can be readily converted to a *para*-bistriflate **44**,[43] suitable for cross-coupling.[44]

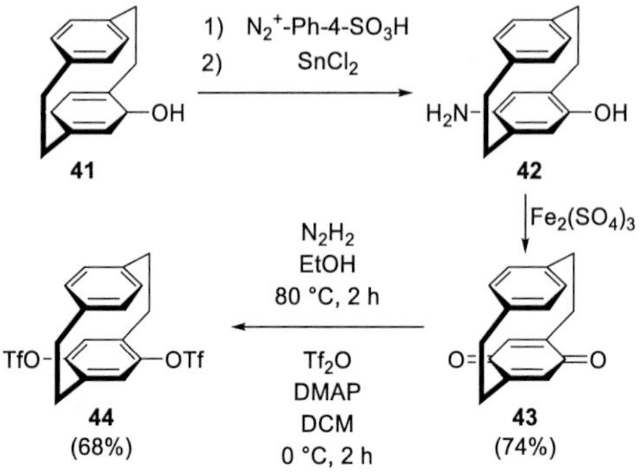

Scheme 15. Oxidative *para*-functionalization with benzoquinone intermediates.

Finally, *para*-bromination was reported for the methoxy derivative **45** in 86% yield but has not gained much attention due to the stability of a methoxy functional group with regard to further derivatization.[45]

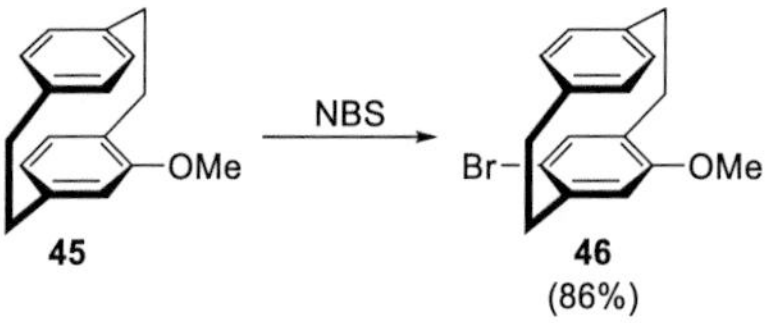

Scheme 16. Direct *para*-bromination of activated 4-methoxy[2.2]paracyclophane (**45**).[45]

1.1.6   Chiral Resolution

Separation of planar chiral derivatives of [2.2]paracyclophane can be a tedious endeavor, especially on a large scale. In order to circumvent expensive chiral (semi)preparative HPLC techniques, chiral resolution *via* derivatizing agents has been reported in some cases. Racemic 4-formyl[2.2]paracyclophane **5** can be easily enantioenriched by repeated fractional crystallization of the diastereomeric mixture of imines **48** (Scheme 17).[46-48] The enantiomerically and diastereomerically pure imine is easily hydrolyzed on silica, yielding (*S*$_P$)-**5** in excellent enantiomeric purity. Although only the (*S*$_P$) isomer is isolated when using the (*R*)-amine **47**, it is possible to obtain the other enantiomer, when (*S*)-**47** is used. The enantiomeric excess is conveniently monitored by ^{1}H NMR of the imine proton. This procedure has proven successful also for an *ortho*-hydroxy-formyl-[2.2]paracyclophane derivative.[47]

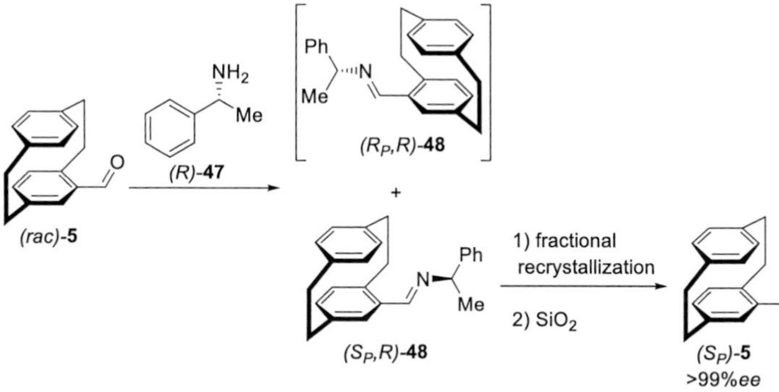

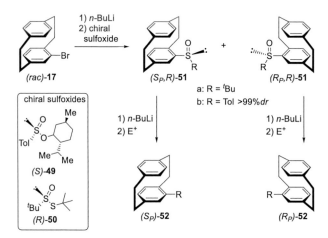

Scheme 17. Chiral resolution of 4-formyl[2.2]paracyclophane **5**.

Ultimately, Rowlands *et al.* reported chiral sulfoxides **51** prepared from thiosulfinate **49**[49] and toluenesulfinate **50**[50] as suitable derivatizing agents for the [2.2]paracyclophane core which can be separated by column chromatography on a 10 gram scale (Scheme 18)[51] with >99%*ee* when **51b** is used.[50] The sulfoxide group can be cleaved by addition of *n*-butyllithium to obtain an enantiopure **Li-52** reactive intermediate which can be conveniently quenched with a number of electrophiles,[49] or can be derivatized to a chiral thiol,[52] essentially promising a single methodology to access a wide range of precursors in an enantiopure way. Furthermore, these chiral sulfoxides are able to direct further functionalization to the *ortho*-position *via* lithiation[49] or pseudo-*geminal* position *via* electrophilic bromination.[53]

Scheme 18. Chiral resolution *via* chiral sulfoxides explored by Rowlands *et al.*[51]

1.1.7   Unprecedented Rearrangements

Due to the structural peculiarities of the [2.2]paracyclophane, this molecule is prone to unexpected reactivity. For instance, a shift of the ethyl bridge has been observed upon addition to the quaternary bridgehead carbon.

The first reported example was observed during a Fischer-type indole synthesis with ethyl pyruvate (**54**) leading to the ene-hydrazine starting material **55** (Scheme 19). An unselective addition to either the quaternary (**56**) or primary (**58**) *ortho*-carbon was observed, which led to the expected fused (4,7)-bridged indolophane **57** in 23% yield and the novel (3,6)-bridged indolophane **59** in identical yield caused by a [1,2]-shift of the ethyl bridge.[54] Even though alkyl-shifts can occur in Fischer-type syntheses,[55] this behavior was only observed when using ethyl pyruvate (**54**) to generate **55**. No such side product was observed when using cyclohexanone and butanone to generate the respective derivatives of **55**. This is explained by an improved enamization of **55** caused by the carboxyethyl group.

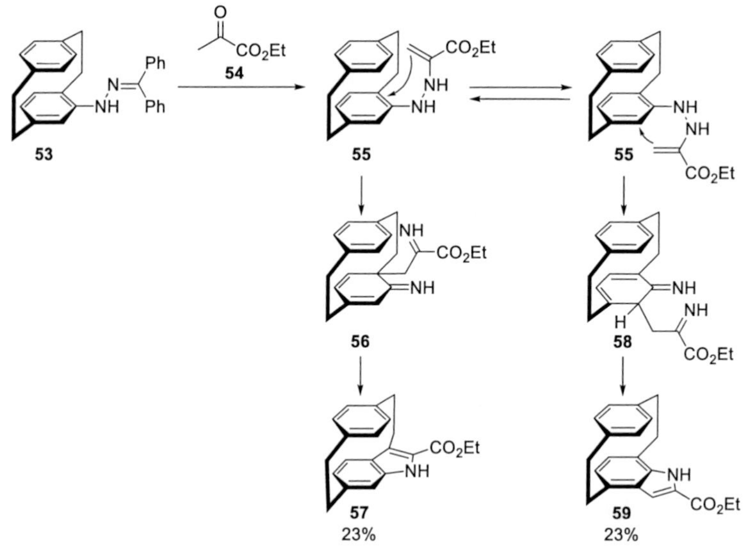

Scheme 19. Fischer-type indole synthesis on [2.2]paracyclophane yielding both typical and atypical indoles **59** and **57**.

A second example for the reactivity of the quaternary bridge-head carbon was observed during a Gould-Jacobs synthesis to access 4-quinolones from the cumulene intermediate **61**.[56] Here, a thermally generated cumulene **61** from decomposition of the malonic enamine precursor **60** proved to generate the expected (5,8)-bridged cyclophane **63** and the unexpected (3,7)-bridged cyclophane **65** in equal amounts. Again a very similar mechanism was proposed starting from a rotamer leading to the addition to the quaternary carbon giving intermediate **64**, which undergoes a subsequent [1,3]-shift.

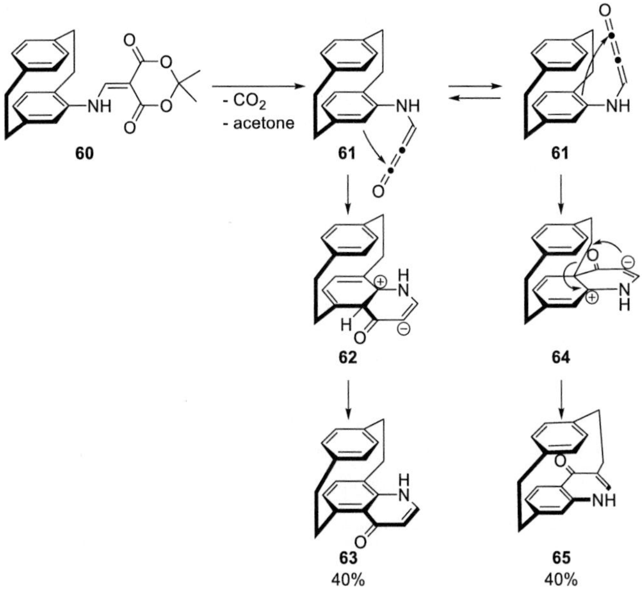

Scheme 20. Gould-Jacobs synthesis of the expected 4-quinolonophane **63** and the unexpected **65**.

In both examples it must be noted, that the isolated yields of the desired and unexpected products were equivalent, essentially demonstrating that there is no selectivity between the primary and quaternary position under the reported conditions. The reactivity of the quaternary carbon under relatively harsh conditions has not been systematically investigated so far, but can give access to a wider range of (hetero)aromatic cyclophanes with short ethyl bridges.

1.2   Chemical Vapor Deposition Polymerization[1]

1.2.1   Concept

First described by Gorham in 1966, [57] [2.2]paracyclophane can be cracked homolytically at the ethylene bridges at high temperatures yielding 1,4-quinodimethanes in gas phase (Scheme 21). In principle this is the reverse process of the [2.2]paracyclophane synthesis described previously in section 1.1.2. The potential of this process was first explored by Gorham in 1966.[57] Under suitable reaction conditions the 1,4-quinodimethane diradicals recombine in a polymeric fashion rather than reform the rigid [2.2]paracyclophane scaffold. Polymer coated surfaces are obtained, indicating that the diradicals in gasphase deposit on the surface prior to recombination. Therefore, this process is coined chemical vapor deposition (CVD) polymerization. The poly-p-xylylenes (PPX) coatings are generally referred to as parylenes. With a suitable device setup and by a carefully tuned gas flux under reduced pressure, CVD can be performed on any object,[58] enabling a generic surface engineering protocol. Although the [2.2]paracyclophane scaffold is cracked under pyrolytic conditions (> 550 °C), numerous functional groups survive this step, tremendously increasing the potential for post-CVD surface engineering, as well.

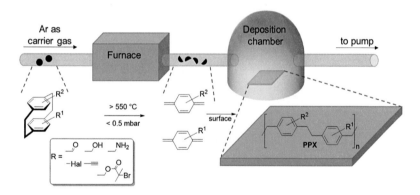

Scheme 21. Mechanism and system setup of chemical vapor deposition (CVD) polymerization.

By a reasonably sophisticated system setup with e.g., numerous pyrolysis chambers or a patterning mask and sequential deposition of different precursors, random copolymers, gradients[59] and microstructuring can be obtained.

This CVD process is especially useful for medical and biological applications as no side products are generated in the homolytic σ-bond dissociation when compared to the alternative access to parylene coatings *via* 1,6-elimination of e.g. chloro-methyl *para*-xylenes yielding hydrochloric acid as a side

---

[1] Parts of this chapter have been published in:
Z. Hassan, E. Spuling, D. M. Knoll, J. Lahann, S. Bräse, *Chem Soc. Rev.* **2018**, *47*, 6947–6963.

product. Although the polymers obtained are not chemically bound to the surface, their high molecular weight and high degree of aromaticity result in mechanically stable (tape test)[60] coatings and often inhibits further characterization due to a lack of solubility. In this regard alkyl substituted [2.2]paracyclophanes were prepared by Greiner and coworkers to permit characterization.[61] The results showed a randomly distributed head to tail selectivity, polymers with a number average of up to 258 000 Da and a polydispersity in the range of 2.6 to 4.5. Although further fundamental research of the Gorham process is still pending the CVD of functionalized [2.2]paracyclophanes yields numerous parylene coatings with intriguing properties and potential applications.

1.2.2   Applications

The chemical inertness and low permittivity make parylenes widely used for coatings of electrical[62] and medical devices.[63-64] The wide range of applications of this process is evaluated, amongst others, by Lahann and coworkers in numerous examples, such as orthogonal immobilization of two different biomolecules by amide formation with surface deposited methylamino groups and Huisgen click reaction to surface deposited terminal alkynes,[65] a pH dependent swelling of the coating when methylamino groups are deposited,[66] and the formation of a backbone degradable polyester copolymer by gas phase ring-opening rearrangement of a cyclic ketene acetal (Scheme 22 top).[67]

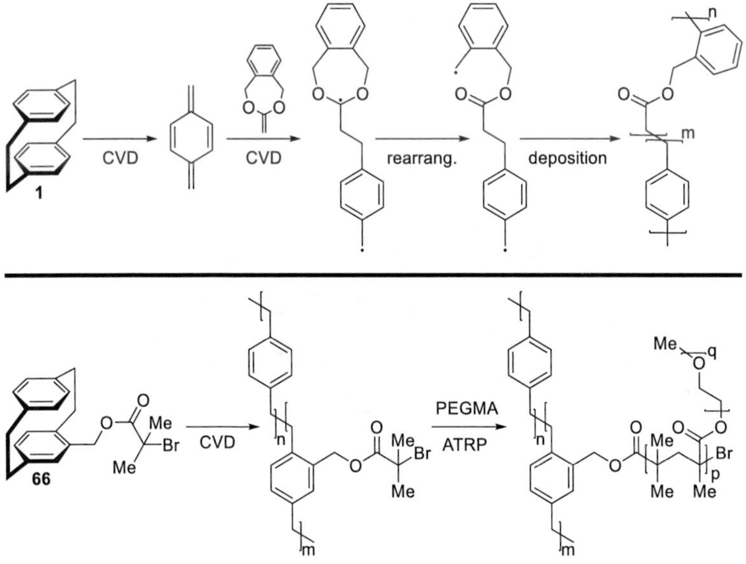

Scheme 22. Top: CVD synthesis of biodegradable parylene copolymers.[67] Bottom: CVD deposition of 2-bromo propyl functionalities gives access to reactive coatings suitable for ATRP polymerization.[68]

In the context of reactive coatings for post CVD functionalization, it has been shown that a suitable ATRP initiator **66** was successfully deposited on surface, enabling a surface-bound yet substrate independent polymerization which was demonstrated with PEGMA (poly(ethylene glycol) methacrylate) in order to obtain bioinert surfaces (Scheme 22 bottom).[68]

Furthermore, CVD was also performed with functional pyridinophane derivatives such as **67** (Scheme 23). In this case polylutidin-co-polyxylylene copolymers were obtained. These polymers show an increased adhesion of HUVEC cells due to the increased polarity and hydophilicity of the material through the incorporated nitrogen.[32] In case of the unsubstituted pyridinophane and the *para*-alkynyl derivative, a patterning and post CVD functionalization was demonstrated by Huisgen click methodology of azido biotin and subsequent incubation with fluorescent streptavidin.

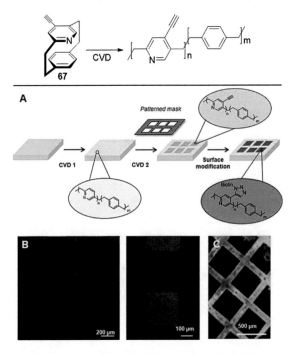

Scheme 23. Top: Formation of functional polylutidine-co-parylene polymers. Bottom: A) Schematic representation and B,C) Demonstration of micropatterning and post CVD functionalization. Reprinted with permission from *Chem. Eur. J.* **2017**, *23*, 13342–13350.[32]

Recently, Lahann *et al.* presented a templated CVD protocol. By using anisotropic liquid crystals (LC), surface anchored nanofibers could be synthesized as schown in Scheme 24.[69] Most notably, by using nematic liquid crystals, linear fibers were obtained, while blue-phase liquid crystals yielded nanostructures with pores of around 500 nm in diameter. For this thesis most relevant is the observation that by using cholesteric liquid crystals, the helical chirality could be reliably transferred to the obtained nanofiber in the form of a spiraling of the fiber in the μm range. Evaluation of adhesion forces clearly showed that interlocking of fibers increases the adhesion force when compared to nontemplated parylene coated surfaces. Additionally, an influence of the relative chirality of two cholesterically templated nanofibers on the adhesion force is observed.

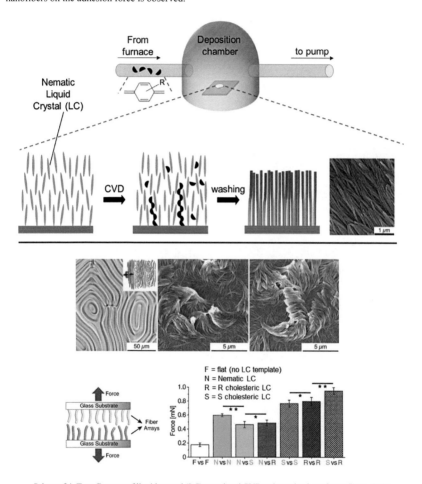

Scheme 24. Top: Concept of liquid crystal (LC) templated CVD polymerization of nanofibers. SEM images of nanofibers. Bottom: Cholesterically templated helical nanofibers with evaluation of adhesion forces. Parts reprinted with permission from *Science* **2018**, *362*, 804–808.[69]

1.3   TADF and OLED

In order to understand the thermally activated delayed fluorescence (TADF) mechanism, and its value for modern light harvesting, the principles of an organic light-emitting diode (OLED) will now be introduced.

1.3.1   Organic Light-Emitting Diodes (OLEDs)

Organic light-emitting diodes are carefully manufactured devices designed to generate light through electroluminescence. This property was first described by Bernanose et al. in 1955 by measuring a mixture of phosphorescent organic dyes in an inert polymer matrix placed in a dielectric cell.[70] Investigation of the basic mechanism only started after this phenomenon was observed on anthracene crystals by Pope et al. in the early 1963.[71-72] The break through report was made in 1987 by Tang and van Slyke who prepared the first dedicated OLED device with an external quantum efficiency of 1% and a brightness of 1000 cd m^{-2} at a driving voltage below 10 V.[73] Therefore, 8-hydroxyquinoline aluminum (Alq$_3$) was deposited in a layered device architecture by vapor deposition, which generated green light (550 nm). The discovery of metal-like conductivity of redox doped polymers by Heeger et al., who were honored with the Nobel Prize in Chemistry in 2000, finally started the industrial development of OLEDs.[74]

Compared to inorganic LEDs, OLEDs offer solution-based processability and the avoidance of precious elements. This promises an inexpensive and a more environmentally friendly fabrication. Furthermore, as the only available planar light source, which can be fabricated onto flexible and transparent substrates, novel design concepts are accessible.

1.3.1.1   Architecture and Functionality of OLEDs

The elementary processes of an OLED device are described in Figure 5. In the first step (I), charge carriers are injected into the organic materials of the OLED device through the cathode and anode. The induced electric field slopes the molecular orbitals. To avoid a spontaneous current flow, it is important that the lowest HOMO is lower than the anode, whereas the highest LUMO needs to be higher than the cathode.[75] Next (step II) the injected charges migrate to their corresponding electrode. As molecules exhibit distinct molecular orbitals rather than extensive valence or conduction bands, used to describe inorganic solid-state matter, the migration of charges is seen as a cascade of redox reactions. This results in a hopping mechanism.[76] The efficiency of conductivity is proportional to the distance, temperature, conformational stability between the reduced and oxidized state and the overlapping of orbitals. Hence π-stacking is an important material property for efficient conductivity in OLEDs.[77] Furthermore, extensively conjugated π-systems are preferred as a decreased energy gap between HOMO and LUMO is obtained.[78] In the third step (III), the migrating positive and negative charges recombine to form so called excitons or excited states located at specific molecules. These molecular excited states undergo relaxation *via* different pathways, of which photoemission (step IV) is the desired process. This process of electrical excitation which relaxes back to the ground state *via* emission of photons is called electroluminescence.[79] The photophysical mechanisms are discussed indepth in the following section 1.3.1.2.

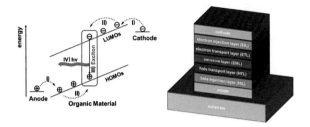

Figure 5. Left) Elementary steps in an OLED device. I) Injection of charge carriers; II) Transport of charge carriers; III) Exciton formation; IV) Light emission. Right) Generalized architectural build-up of OLEDs.

The actual build-up of an OLED device in order obtain a most efficient generation of light is quite sophisticated, as numerous layers need to exhibit distinct features in order to facilitate the charge migration and increase accumulation at the emissive layer. A generalized OLED architecture is presented in Figure 5. The respective purposes are charge injection from the corresponding electrodes, charge transport and blocking and out-coupling of the generated light.[80] The key characteristics of the layer materials are their HOMO or LUMO levels which smoothen the charge transport towards the final emissive layer (EML).

Substrate

In principle the substrate is only needed as the carrier material for the whole OLED architecture. In order to out-couple light, this layer needs to be transparent and depending on the application it can also be flexible. Therefore glass, or transparent polymers are used.[75]

Anode and Hole Injection Layer (HIL)

As cathodes are made of base metals or alloys which are opaque, the anode has to be transparent to allow the outcoupling of light. Indium tin oxide (ITO) $(In_2O_3)_{0.9}(SnO_2)_{0.1}$ has been established as the standard material for this purpose. To avoid indium, other transparent conductive oxides (TCO),[81] or purely organic anodes are explored.[82] As the interfaces of the anode layers are typically rough, a smoothening hole injection layer (HIL) is often deposited on top to level them. Furthermore, the transfer to the following hole transport layer (HTL) can be disposed.

The HIL material must have an electron affinity in the range between the anode and the following layer alleviating the hopping mechanism of the cationic charges *via* HOMOs.[75] The so called PEDOT:PSS material is commonly used for this purpose. It is a mixture of a conjugated polythiophen polymer and a sulfonic acid derivative of polystyrene.

Hole Transport Layer (HTL) / Electron Blocking Layer (EBL)

The purpose of the HTL is to guide and smoothen the transport of positive charges *via* HOMOs towards the final emissive layer (EML). Additionally, since the charges need to be accumulated within the emissive layer, this layer functions as an electron blocking layer (EBL) for the anionic charges migrating from the cathode. These properties are realized by electron rich heteroaromatic compounds.[75]

Emissive Layer (EML)

In the emissive layer, the charges meet to form excitons. This recombination of opposite charges has to be even to avoid a build-up of charge, as this would induce detrimental and irreversible redox reactions within the material. This layer is doped with electroluminescent emitters, which are responsible to the generation of light *via* photophysical processes discussed below in section 1.3.1.2. The concentration of electroluminescent emitters must not be too high to exclude intramolecular effects such as triplet-triplet-annihilation, but the emissive molecules need to act as isolated sites for light generation within the emissive layer as the host material. This material must be inert to the redox and photophysical conditions, and be able to bring both positive and negative charges to the emitters.

Electron Transport Layer (ETL) / Hole Blocking Layer (HBL)

The electron transport layer (ETL) has the identical purpose, as the HTL. However, it is related to the electron transfer *via* LUMO hopping. Suitable materials commonly exhibit electron deficient aromatic compounds such as triazole, imidazoles or pyridines.[75, 78] As this layer is also used to block positive charges from leaving the emissive layer, it can also be called a hole blocking layer (HBL).

Electron Injection Layer (EIL) and Cathode

Again, the electron injection layer (EIL) has the same purpose as the corresponding HIL, but with regard to negative charges transported *via* LUMOs. For this purpose, mixtures of phenanthroline derivatives and quinodimethane derivatives can be used.[83] As the anode is commonly the transparent side of the OLED stack, cheap anode materials such as magnesium, aluminum or alloys of lithium are used.

Fabrication Techniques

Tang and van Slyke used vapor deposition techniques to generate the first OLED.[73] This process offers superior purity of the device as all materials used must be sublimable and can therefore be thoroughly purified by sublimation before the actual device fabrication, which hampers the application of polymers. Another disadvantage is that high-vacuum deposition chambers are costly to maintain especially with panels of increasing size and the fabrication process is batch-based.

Alternatively, solution-based fabrication techniques are explored,[84] as they can offer continuous fabrication processes *via* printing techniques, which are suitable for non-sublimable materials such as heavy emitters or polymers, but a very selective solubility must be present to avoid mixing of layers. To address this issue, cross-linking techniques have been developed.[79, 85]

Ultimately, the operating conditions of OLEDs are harsh as the materials are constantly brought to an oxidized or excited state. For this purpose, reliable encapsulation must be ensured, which is especially challenging if flexible substrates are used.[78]

Out-Coupling of Light

Finally, the generated photon waves propagate in every direction. Only the photons that manage to out-couple through the layers leaving *via* the transparent anode can be seen. The external quantum efficiency (EQE) quantifies the percentage of useable photons.

The multi-layered setup of the device forces the photons to pass several interfaces which exhibit different refractive indices. This can cause refraction and trap the photons within a layer.[86] Specific interface structuring is investigated to enhance the out-coupling by avoiding the refraction.[87]

1.3.1.2   Photophysics of OLEDs

In contrast to excitation *via* photo absorption which selectively yields singlet excited states, electroluminescence is a redox-based process which randomly produces singlet (S_1) and triplet (T_1) excited states. As there are three triplet states and only one excited singlet state, the distribution is 25% *versus* 75% (Scheme 25).

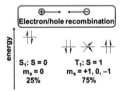

Scheme 25. Quantum statistics state that the multiplicities of the angular momenta m_s form a 25/75 ratio between excited singlet states and triplet states due to the random nature of electron/hole recombination.

As fluorescence is only allowed from the excited singlet state S_1 to the ground state S_0. The maximum quantum efficiency possible is 25% for purely fluorescent compounds. The Jablonski diagram for organic compounds is shown in Scheme 26. Additionally, due to Hund's rule and the spin-orbit coupling, the intersystem crossing (ISC) turns into a competitive mechanism and thus lowers the fluorescence yield.[88] Furthermore, nonradiative relaxation mechanisms (e.g. *via* exciton-phonon coupling[89]), refraction at interfaces and many other losses add up in device, making this system very inefficient.[75]

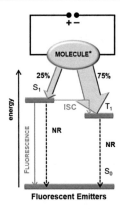

Fluorescent Emitters

Scheme 26. Jablonski diagram for purely fluorescent emitters. Electric circuit brings a molecule to
its excited state indicated with *. The basic pathways of relaxation after electrical excitation include
fluorescence or nonradiative relaxation (NR) and intersystem crossing (ISC).

The first generation of OLEDs only using fluorescence to generate light, was soon succeeded by the
second generation of OLEDs, where promoted phosphorescence was employed to raise the internal
quantum efficiency. As heavy atoms such as iridium or platinum weaken the selection rules by an
enhanced spin-orbit coupling, ISC and phosphorescence ($T_1 \rightarrow S_0$) become enabled relaxation pathways
as seen in Scheme 27, and the respective emitters are therefore also coined triplet harvesters.[90] In
contrast to purely fluorescent emitters, this generation allows to up to 100% of the generated excitons
for light generation. Unfortunately, the need of heavy and therefore rare elements makes their use in
large scale application expensive.

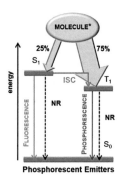

Phosphorescent Emitters

Scheme 27. Jablonski diagram with promoted phosphorescence.

1.3.2   Thermally Activated Delayed Fluorescence (TADF)

The most recent third generation of OLED emitters is able to use all of the available excitons to harvest light as well, but without the need of heavy metal complexes and therefore received much attention in OLED research following the seminal works of Adachi *et al.*[91-94] The so called thermally activated delayed fluorescence (TADF) is now regarded as the most promising mechanism for harvesting excitons[95] in electroluminescent devices. In this concept, the excited S_1 and T_1 states are placed to minimize the energy gap (ΔE_{ST}) (Scheme 28). This permits a rapid equilibration *via* reversed intersystem crossing (RISC) from spin-orbit coupling of isoenergetic levels of S_1 and T_1. Although this mechanism had been known for aromatic ketones, xanthene derivatives and organo-transition metal compounds for a long time,[96-100] it was first considered in OLED research in 2009 by Adachi *et al.* for tin(IV)porphyrin complexes.[91] As OLEDs are operated at ambient temperatures this gap must be small enough to allow efficient RISC at ambient temperature. Adachi *et al.* proposed a ≤ 100 meV criteria for the energetic gap of promising TADF molecules.[93] In principle, it is possible to claim that environmental thermal energy is harvested to generate light.[92]

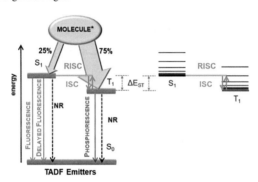

Scheme 28. Jablonski diagram for emitters employing the thermally activated delayed fluorescence (TADF) principle with the ISC/RISC (intersystem crossing and reverse-ISC) emphasized.

Since fluorescence is a considerably faster process than phosphorescence, all excitons can be harvested by (delayed) fluorescent emission. With RISC as the rate determining step, which is only permitted at elevated temperatures, the turn-on of delayed fluorescence (or absence of phosphorescence) can be followed by time-dependant emission decay time measurements (Figure 6).[95, 101]

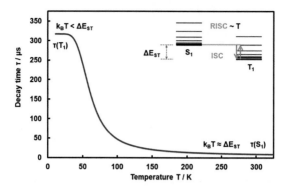

Figure 6. The dependence of the emission decay time τ on the temperature T as proof of TADF. Long decay times in the range of ms indicate an emission *via* phosphorescence, while short decay times (μs) indicate (delayed) fluorescence.[88]

A small ΔE_{ST} is obtained *via* a minute exchange integral between the HOMO and LUMO, as was described by Adachi *et al*.[93, 102] However this can be accompanied by a reduced oscillator strength *f*, which ultimately leads to compromised photoluminescent quantum yield (Φ_{PL}). Therefore, strict design guidelines must be followed to maintain the onset of TADF and the brightness of the emitters (Figure 7).[103] In TADF terminology, the functional moieties containing the HOMO are called donors, while the groups containing the LUMO are named acceptors. These two functional groups are attached to a steric hindrance structure. The vast majority of reported emitters follow a twist principle in order to reduce the overlap of the frontier molecular orbitals (FMOs).[88, 95]

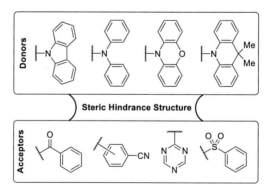

Figure 7. Generic TADF donor and acceptor structures.[88]

A well-known example is the **4CzIPN** emitter (**68**) published by Adachi *et al.*,[93] which shows the reduced overlap *via* concentration of the HOMO mainly on the four carbazole units, while the nitriles of the isophthalonitrile subunit contain the LUMO. Both orbitals are oriented perpendicularly to each other.

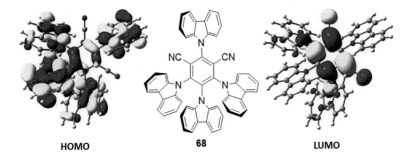

HOMO   **68**   **LUMO**

Figure 8. HOMOs are located at the carbazolyl moieties, LUMOs are located at the nitriles. Images taken from Adachi *et al.*[93]

As this research field is very active, recent findings indicate that planar structures are able to exhibit TADF as well, while maintaining a strong oscillator strength.[104-105] This example shows that the details of the TADF mechanism are still under investigation.[106]

Efficient blue emitters are considered as the great goal of OLED research. Molecular TADF design not only relies on large torsions, but also on weak donor units to deepen the HOMO or on weak acceptor units to raise the LUMO. Unfortunately, when both functional moieties are weak, then the frontier molecular orbitals are too dispersed to assure a reduced FMO overlap. A careful balance must be found to obtain performant TADF emitters in the blue region.

1.3.2.1   Through-Space Design

A less explored, but striking design concept for TADF that has been investigated mainly by Swager, Baldo *et al.* is the electronic communication of donor and acceptor groups mediated by through-space conjugation. This excludes spiro-linked emitters which allow an electronic communication *via* one sp^3-center. So far, a triptycene skeleton confining the donor and acceptor units in a 120° orientation,[107] and a xanthene-linked cofacial orientation of the donor and acceptor with a distance of 3.3–3.5 Å were reported (Figure 9).[108] In case of the triptycene emitters **69** and **70** a τ_D of 6.5 μs and 2.4 μs was observed, respectively, which is in accordance with the small ΔE_{ST} values calculated. As the oscillator strength (*f*) is calculated without considering through-space interaction, it can be taken as an indicator of how exclusive this through-space interaction is with respect to luminance. OLED devices were fabricated, showing EQE_{max} values of 4.0% and 9.4% in the green (542 nm) and yellow (573 nm) range. Six donor-expanded derivatives of **69** were reported by Geng, Su *et al.* in silico, with the smallest ΔE_{ST} of 26 meV for a *tert*-butylcarbazole derivative.

Following the xanthene-bridged donor/acceptor design, emitters **71**–**73** were reported, with even smaller oscillator strengths of 0.0001-0.008, yet accompanied by minute ΔE_{ST} values of 0.1–8 meV. As the carbazole emitter **72** showed reduced stability, only OLEDs with the phenoxazine donor **71** and the *tert*-butylcarbazole donor **73** were fabricated with again moderate EQE_{max} values of 10% and 4% in the yellow (584 nm) and sky-blue (488 nm) spectrum. Although TADF was observed with delayed lifetimes, τ_D of 3.3 μs and 4.1 μs accompanied by moderate Φ_{PL}, the through-space electronic communication mainly occurred *via* C–H··π interactions, creating efficient aggregation induced emission.

As the design of a through-space mediated TADF emitter based on a [2.2]paracyclophane core was a main target of this thesis (*vide infra*), the results of a competing group are mentioned in Figure 9. The results of that group were published after the completion and publication of excerpts of this thesis.[109] Recently, Zhao *et al.*[110] reported on [2.2]paracyclophane-based through-space mediated TADF emitters with a pseudo-*geminal* and a pseudo-*meta* orientation of the dimethylamino donor and dimesitylboron acceptor yielding **74** and **75**. This design gave increased ΔE_{ST} values compared to the xanthene and triptycene design, but increased through bond oscillator strengths in the range of **70**. As the reported τ_D values of 0.38 μs and 0.22 μs for emitters **74** and **75**, respectively, are questionably low in order to claim the turn-on of TADF, no devices were fabricated. But chiral separation of the racemic emitter **74** showed distinct CPL emission for each enantiomer with a maximum g_{lum} value of 4.24 × 10^{-3}.

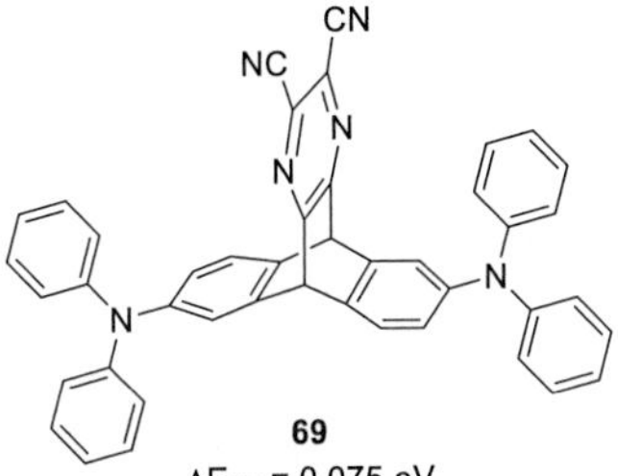

69
ΔE_{ST} = 0.075 eV
f = 0.008
λ_{EL} = 542 nm
EQE_{max} = 4.0%

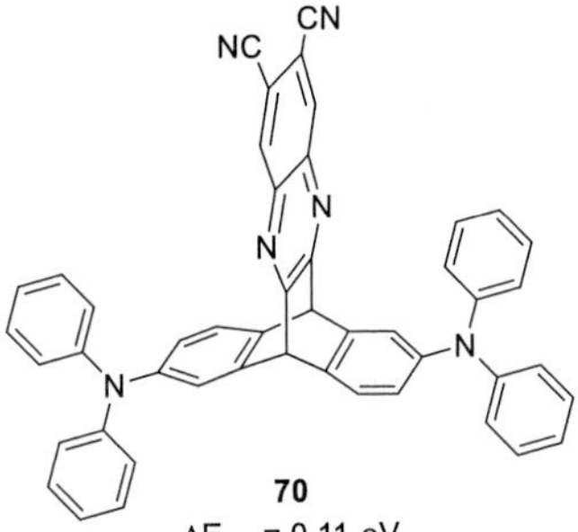

70
ΔE_{ST} = 0.11 eV
f = 0.084
λ_{EL} = 573 nm
EQE_{max} = 9.4%

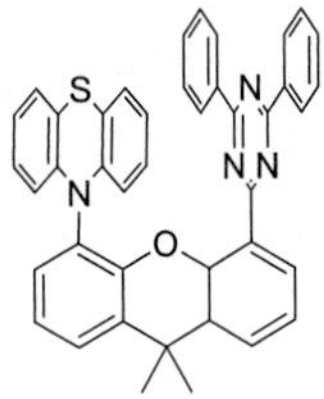

71
ΔE_{ST} = 0.0001 eV
f = 0.00007
λ_{EL} = 584 nm
EQE_{max} = 10%

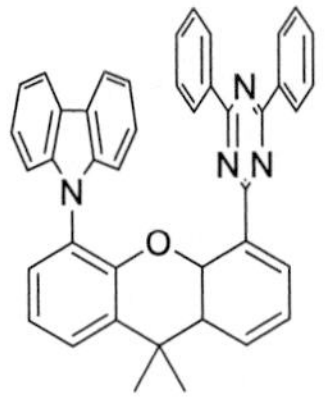

72
ΔE_{ST} = 0.008 eV
f = 0.0003
λ_{PL} = 418 nm

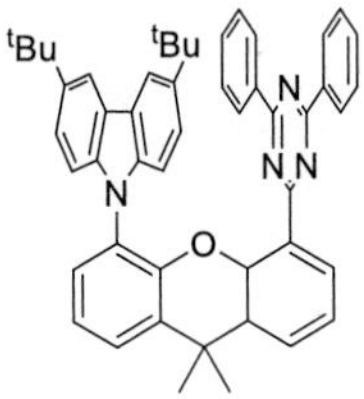

73
ΔE_{ST} = 0.007 eV
f = 0.0005
λ_{EL} = 488 nm
EQE_{max} = 4%

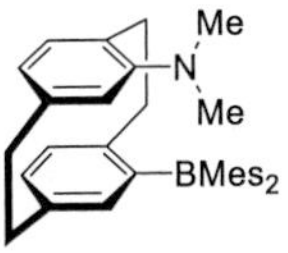

74
ΔE_{ST} = 0.17 eV
f = 0.054
λ_{PL} = 425 nm
g_{lum} = 4.24 x 10^{-3}

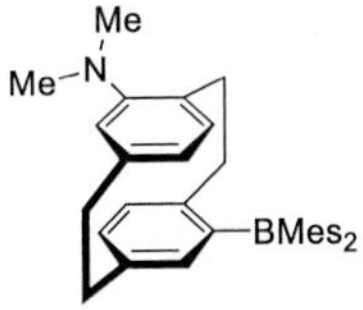

75
ΔE_{ST} = 0.12 eV
f = 0.0209
λ_{PL} = 487 nm

Figure 9. Overview of reported through-space TADF emitters.[107-108, 110]

1.3.2.2   Triazine Twist Design

The turn-on of TADF *via* twisting can be exemplified with the TADF research conducted on the simple *N*-carbazolyl-4-triphenyltriazine D–A structure **76** (Figure 10). This structural class commonly does not display low oscillator strengths. As the parent compound shows deep-blue photoluminescent emission of 381 nm, which is a key goal of OLED research, it does not exhibit TADF due to a HOMO–LUMO splitting of 0.30 eV, which does not meet the criteria for an efficient RISC process.[111] Early attempts to optimize this structure were conduction by substitution of the donor unit to phenoxazine yielding **77**,[112] or dimethylacridine yielding **78**.[113] A significant decrease of ΔE_{ST} to 0.07 and 0.05 meV was reported, accompanied by a significant increase of the torsion angle to 74.9° and 88° for the respective emitters. This can be explained by the increased proximity of the *ortho*-hydrogen atoms caused by the increased size of the six-membered 1,4-oxazine and 1,4-(4,4'-dimethyl)dihydropyridine subunits compared to the five-membered pyrrole subunit in the carbazole donor of **76**. The change of structure, conformation and ultimately the energetic profile of emitters **77** and **78** resulted in TADF turn-on proven by device performances (EQE$_{max}$) of 12.5% and 26.6%, although a large red-shift to blue-green (500 nm) and green (529 nm) was observed. To stay in the blue region, clearly the weak carbazole donor unit was to be kept, but an increase of twist was needed. To meet these criteria, emitters with methyl groups at the *ortho*-positions to the N–C bond were introduced by Adachi *et al.* to yield emitters **79**–**81**.[114] All emitters showed a distinct torsion angle of 71.3°–86.7°, ΔE_{ST} of 0.08–0.17 eV and excellent device performances (EQE$_{max}$) of 18–22%. Only emitters **80** and **81** managed to give deep-blue electroluminescence of 450 nm and 452 nm, respectively, while emitter **80** showed a significant red-shift to 500 nm. This can be explained by the dramatically reduced ionization potential of the 1,3,6,8-tetramethylcarbazole, making the triplet state a stable ^3CT (charge transfer) state rather than a ^3LE (locally excited) state. Swager *et al.* introduced a triptycenylcarbazole unit to determine if the HOMO–LUMO splitting can be reduced effectively without lowering the S_1 state needed for deep-blue emission *via* homoconjugation.[115] The resulting emitters **82** and **83** showed decreased torsion angles of 51.6° and 69.5° and worse device performances of (EQE$_{max}$) 10.4% and 11%, respectively when compared to emitters **79**–**81**. When emitter **82** is compared to the parent compound **76**, a TADF turn-on could be achieved by expansion of the HOMO without introducing additional torsion to the system with a delayed emission lifetime (τ_d) of 38 µs for **82** and no detected delayed emission component for **76**. Finally, emitter **83** proves that additional torsion is the most straight forward method to reduce ΔE_{ST}.

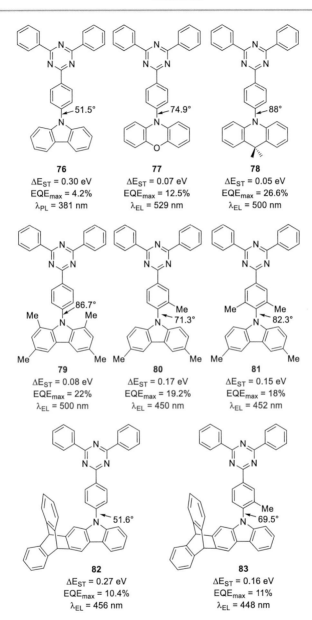

76
ΔE_{ST} = 0.30 eV
EQE_{max} = 4.2%
λ_{PL} = 381 nm

77
ΔE_{ST} = 0.07 eV
EQE_{max} = 12.5%
λ_{EL} = 529 nm

78
ΔE_{ST} = 0.05 eV
EQE_{max} = 26.6%
λ_{EL} = 500 nm

79
ΔE_{ST} = 0.08 eV
EQE_{max} = 22%
λ_{EL} = 500 nm

80
ΔE_{ST} = 0.17 eV
EQE_{max} = 19.2%
λ_{EL} = 450 nm

81
ΔE_{ST} = 0.15 eV
EQE_{max} = 18%
λ_{EL} = 452 nm

82
ΔE_{ST} = 0.27 eV
EQE_{max} = 10.4%
λ_{EL} = 456 nm

83
ΔE_{ST} = 0.16 eV
EQE_{max} = 11%
λ_{EL} = 448 nm

Figure 10. Overview of published examples of TADF turn-on of the parent triazine emitter **76**. All values are given from the reported device architectures and all torsion angles are reported from DFT calculations.[111-115]

2   Objective

Since the first discovery of [2.2]paracyclophane (**1**) by Brown and Farthing exactly 70 years ago[1] as a byproduct of the pyrolysis of *para*-xylene,[26] this molecule has sparked tremendous interest in the scientific community. The two co-planar phenyl rings in close proximity of 3.09 Å,[2-3] fixed by two short ethyl bridge units in *para*-position result in a "*bent and battered*"[2] molecular skeleton which is unknown in conventional non-cyclophane aromatic chemistry. The key features of this molecule are I) the unique electronic situation enabling transannular electronic communication of the two π-systems,[4-5] II) the presence of planar chirality upon substitution and III) the considerable spherical bulk induced by the adjacent ethyl bridge and second coplanar phenyl ring. Despite these unique features, the bent and therefore unfavored conformation influences the chemistry of this molecule dramatically, making the synthetic access to desired molecules a challenging endeavor.

Up until now the [2.2]paracyclophane scaffold has predominantly been investigated in catalysis as a chiral ligand,[7-9] exploiting the chirality and steric hindrance of this molecule. An example of this is the commercialization of the chiral [2.2]paracyclophane-based phanephos ligand.[116] In contrast, investigations regarding material sciences are far less explored.[6]

Therefore, the aim of this thesis was to establish a synthetic access to [2.2]paracyclophane-based derivatives for different applications in material sciences and supramolecular chemistry (Figure 11).

In the first project, thermally activated delayed fluorescence (TADF) emitters are synthesized with the [2.2]paracyclophane serving as a bridge between an electron rich and an electron poor aromatic substituent to mediate through-space electronic communication. For this, a synthetic access to pseudo-*geminal* and pseudo-*para* oriented functional groups has to be developed. Since no general cross-coupling protocol is known, and steric hindrance for the pseudo-*geminal* emitter group is expected, these syntheses must be carefully optimized.

Additionally, a second class of TADF emitters is designed and synthesized using the [2.2]paracyclophane as an extension of the electron rich (1,4)carbazolophane donor group in combination with a triazine acceptor unit to turn-on the TADF mechanism. Furthermore, the chirality of the resulting emitter which is introduced by the substituted [2.2]paracyclophane scaffold should be evaluated through a stereoselective access to this emitter.

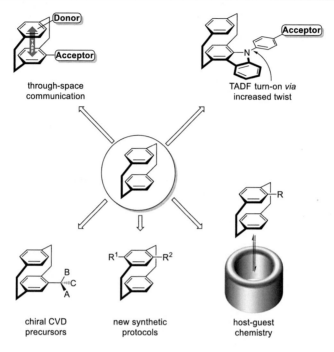

through-space
communication

TADF turn-on *via*
increased twist

chiral CVD
precursors

new synthetic
protocols

host-guest
chemistry

Figure 11. Objectives of this thesis.

In a second project, new [2.2]paracyclophane-based chemical vapor deposition (CVD) monomers are synthesized in order to tackle the recently posed question, whether it is possible to transfer chirality from a CVD monomer to a parylene polymer, for this the access to enantiopure derivatives is be explored.

Finally, the assumption that certain substituted [2.2]paracyclophanes can be promising guest molecules in supramolecular chemistry with cucurbit[8]uril has to be verified by evaluation of a suitable synthetic access to 4-(4'-pyridyl) derivatives.

3 Main Section

3.1 Through-Space TADF Emitter Design[2]

Although TADF is a rapidly growing research area, the through-space design principle is little explored (see chapter 1.3.2.1 in the introduction). Apart from the reported triptycene[107] and xanthene[108] scaffolds, the [2.2]paracyclophane is a well suited scaffold for stronger transannular electronic communication between the benzene decks. The reported deck distance of 3.09 Å,[2] which is smaller than the van der Waals distance between layers of graphite (3.35 Å) has sparked numerous investigations focusing on π–π transannular interaction in the last decades. Bazan *et al.* investigated this *via* D–A stilbene derivatives, which showed nonlinear optical properties and considerable through-space charge transfer,[4-5] including strong positive solvatochromism caused by a polarizable electronic structure in the excited state.[117] Morisaki, Chujo *et al.* focused on [2.2]paracyclophane-based through-space π-extended conjugated polymers. It was demonstrated that electronic interactions can be observed through more than ten [2.2]paracyclophane-bridged conjugation layers, yet emission in these systems occurs from the isolated monomer π systems, giving access to well-defined monomer-localized HOMO-LUMO gaps rather than a broad valence-conduction band gap in the polymers.[118-119] Additionally, given the inherent planar chirality in functionalized [2.2]paracyclophanes, this scaffold can yield chromophores with intense circular polarized luminescence with an astonishingly high g_{lum} of 10^{-2} which is rare for hydrocarbons.[120-123] In spite of these fascinating properties, [2.2]paracyclophane chemistry suffers from challenging or sometimes unpredictable reactivity. Although various successful cross-coupling protocols have been reported,[44] a direct C–N coupling to an *N*-heterocycle such as carbazole or diphenylamine, though claimed in patents,[124-125] has never been reported in the scientific literature so far.

Therefore, the aim of this project was to synthesize through-space conjugated TADF emitters which exploit the unique electronic transannular interaction of the [2.2]paracyclophane core (Figure 12).

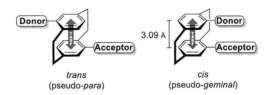

Figure 12. Design concept of [2.2]paracyclophane mediated through-space conjugated TADF emitters.

---

[2] Parts of this chapter have been published in:
E. Spuling, N. Sharma, I. D. W. Samuel, E. Zysman-Colman, S. Bräse, *Chem. Commun.* **2018**, *54*, 9278–9281.

As numerous configuration isomers are possible, it was chosen to limit the potential compound scope to pseudo-*geminal* (also called *cis*) and pseudo-*para* (also called *trans*), as these isomers are expected to give the most pronounced distinction of photophysical properties, if any occur.

3.1.1   Retrosynthetic Analysis

For each isomer, a separate synthetic route was determined. As the donor units are usually *N*-coupled aromatic derivatives and the acceptor units are electron withdrawing groups, such as sulfones, ketones or nitriles (see intro) a suitable synthetic path was to be determined. For this purpose, a sequence of I) acquiring the correct relative orientation, II) conversion to a bromo-acceptor intermediate, and III) cross-coupling with the donor building block to the final emitter; was conducted.

The pseudo-*geminal* orientation cannot be obtained as easily as the following pseudo-*para* orientation. To get this configuration, the transannular directing effect (see chapter 1.1.3) of carbonyl derivatives has to be exploited in the first step of the synthetic path (Scheme 29). Subsequently, the carbonyl group has to be transferred to the desired acceptor group with a bromide in the pseudo-*geminal* position. In the final step a cross-coupling similar to the procedure for the pseudo-*para* isomer route yields the final emitter.

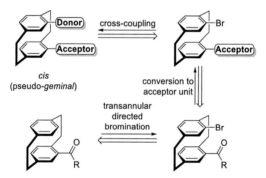

Scheme 29. Retrosynthetic analysis of pseudo-*geminal* TADF systems.

In case of the *trans* (or pseudo-*para*) isomer (Scheme 30), the correct relative orientation of the two substituent could be obtained by fractional crystallization of the insoluble pseudo-*para* dibromide **22**. From this intermediate, a monofunctionalization *via* monolithiation with subsequent electrophilic quenching can be achieved, followed by conversion to the desired acceptor unit. In the final step, the residual bromo substituent can be targeted by cross-coupling techniques to form conjugated C–C (Suzuki-Miyaura, Stille, Negishi etc.) or C–N (Hartwig-Buchwald) bonds to suitable donor units.

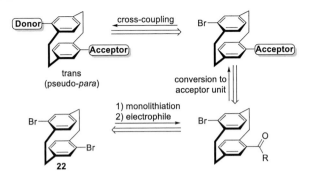

Scheme 30. Retrosynthetic analysis of pseudo-*para* TADF systems.

3.1.1.1   Previous Results

The first attempt to a pseudo-*geminal* benzoyl carbazolyl/diphenylamino emitter during my MSc thesis showed that the bromide is a challenging substrate for cross-coupling (Scheme 31).[126] This is further underlined by the lack of reported C–N coupling of diarylamino groups to the [2.2]paracyclophane core. During that work, a key focus laid on the synthetic approach to donor functionalization on an acceptor-free model system. It was shown that both direct Hartwig-Buchwald and Ullmann couplings of carbazole and diphenylamine to the 4-bromo[2.2]paracyclophane (**17**) and even an indirect Hartwig-Buchwald coupling of aniline to *N*-phenylamino[2.2]paracyclophane (**36**) turned out to be unsuccessful.

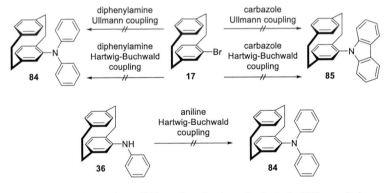

Scheme 31. Previous results on C–N coupling of a donor directly to the [2.2]paracyclophane core.[126]

These results were taken into account for the synthetic feasibility of the desired TADF groups.

3.1.1.2   Choice of Donor and Acceptor Units

Due to these negative results, a C–C Suzuki-Miyaura cross-coupling was chosen, as it proved successful for 4-pyridyl coupling in another project (see chapter 3.5),[44] and was further optimized in the Bräse group.[127] This diversion meant that a phenyl linker had to be incorporated into the TADF design. The benzoyl and the nitrile groups were chosen as acceptors, since the benzoyl is synthetically easily available, and the nitrile is a strongly electron withdrawing acceptor group deepening the LUMO level and contracting the LUMO orbital distribution to the cyanoarene.[103] Additionally, the nitrile provides a synthetic access to weaker acceptors such as oxadiazoles.[128] Additionally, the LUMO of the cyanoarene moiety will be undoubtedly distributed on the [2.2]paracyclophane scaffold. An overview of the intended donors and acceptors are given in Figure 13.

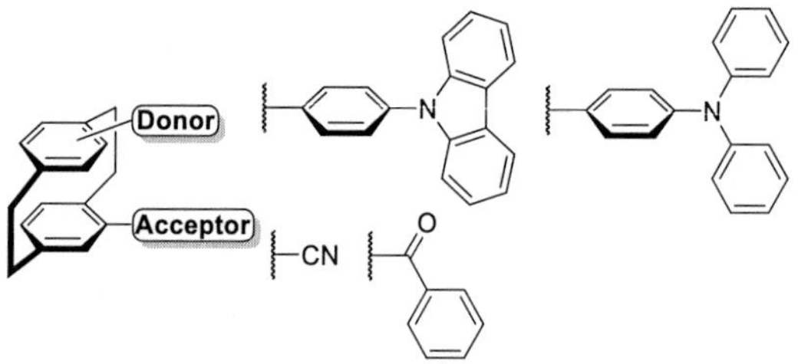

Figure 13. Overview of the used donors and acceptors groups.

As the retrosynthetic analysis showed, the synthesis of the respective key intermediates with the finalized acceptor group and a bromide in the correct relative configuration proceeds *via* initial construction of the acceptor group. Therefore, the subchapters are divided into benzoyl and nitrile approaches.

3.1.2   Benzoyl Acceptor for TADF

3.1.2.1   Synthesis of pseudo-*geminal* Emitters

At first the pseudo-*geminal* emitters based on a benzoyl acceptor were synthesized. The key step is the transannular directed bromination to the pseudo-*geminal* position from a suitable directing group (see Scheme 29), which is commonly a carbonyl structural unit as present in ketones, esters, and amides. Gratifingly, the benzoyl group could be used to direct the bromination to the pseudo-*geminal* position while being a complete TADF acceptor group. To get there the [2.2]paracyclophane (**1**) was benzoylated *via* Friedel-Crafts acylation using benzoyl chloride and aluminium chloride as Lewis acid. The desired 4-benzoyl[2.2]paracyclophane (**86**) was obtained in 79% yield as depicted in Scheme 32.

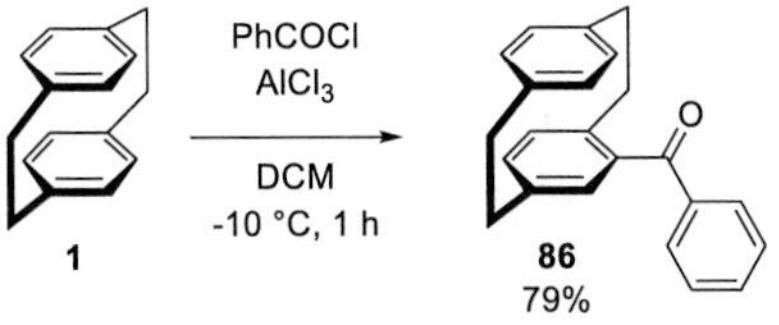

Scheme 32. Synthesis of benzoylated [2.2]paracyclophane **86**.

Next, iron-catalyzed electrophilic bromination of **86** was conducted yielding the pseudo-*geminal* bromobenzoyl derivative **87** in 46% isolated yield, but excellent selectivity (Scheme 33). No other regioisomer was observed.

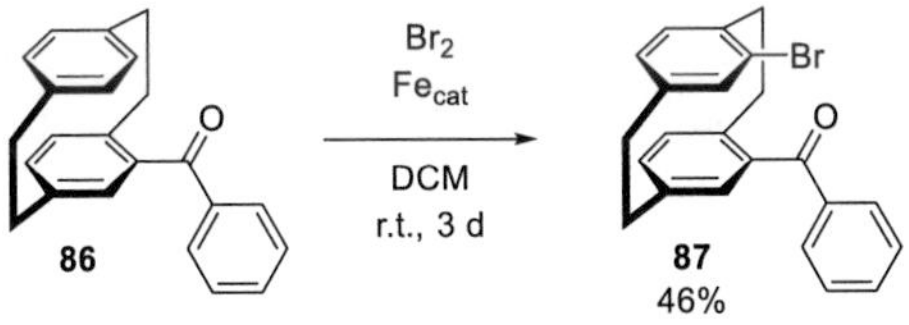

Scheme 33. Transannular directed electrophilic bromination to the pseudo-*geminal* substituted bromo benzoyl derivative **87**.

Furthermore, the absolute configuration could be verified by crystallography (Figure 14), showing that indeed the pseudo-*geminal* orientation of isomers was obtained. When the deck distances of **87** and the reported values for the [2.2]paracyclophane (**1**) were compared, [2-3] the distance of 2.76 Å at the bridgehead carbon atoms turned out to be nearly identical to the literature value of 2.78 Å. In contrast, the distances between the substituent-connected carbon atoms at the paracyclophane decks measures 3.18 Å, which is more than the 3.09 Å reported for the unsubstituted [2.2]paracyclophane. Further analysis shows that the bromine atom and the carbonyl carbon both bend out of the respective deck planes, leaving a distance of 3.53 Å. The distance of the bromine to the carbonyl oxygen atom amounts to 3.34 Å, and the distance to the closest ethyl bridge hydrogen atom is 2.90 Å. This brings the bromine in a relatively encumbered position, which might be problematic for a cross-coupling procedure, where a bulky active complex has to be formed.

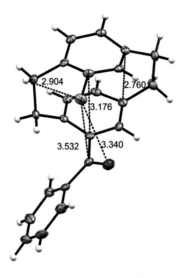

Figure 14. Molecular structure of pseudo-*geminal* substituted **87** drawn at 50% probability level.

The final cross-coupling was screened (Table 1) using an excess of two equivalents of 4-(*N*-carbazolyl)phenyl boronic acid (**88**). Initially, a self-optimized Suzuki-Miyaura protocol for 4-pyridyl boronic acids (see chapter 3.5)[44] was tested (entry 1) giving solely a dehalogenated product. Although dehalogenation is commonly observed when alcohols are used as solvents due to β-hydride elimination,[129], water was presumed to protonate the oxidative Pd complex with help of the adjacent ketone. Furthermore, elevated temperature can promote dehalogenation as well. Therefore, water was excluded by switching to DMF as a water-free solvent. This was accompanied by an increased base loading of 4.00 equivalents of K_3PO_4 (entry 2), which still gave dehalogenated product **86**. Further decrease of the temperature to 50 °C did not improve the reaction outcome (entry 3), but it still gave considerable amounts of dehalogenated product. A switch to the $Pd(OAc)_2$/RuPhos catalyst system, which was recently reported as a highly active Suzuki-Miyaura catalyst system for a [2.2]paracyclophane trifluoroborate,[127] finally gave the desired product in 5% yield (entry 4), although in this work, the reactivities are switched compared to the literature. Doubling the catalyst loading to 10 mol% $Pd(OAc)_2$ / 20 mol% RuPhos yielded the emitter in 57%, which was sufficient to obtain reasonable amounts of the through-space conjugated emitter **89**. Unfortunately, the isolated compound showed to be nonluminescent.

Table 1. Screening of Suzuki-Miyaura cross-coupling conditions to emitter **89**.

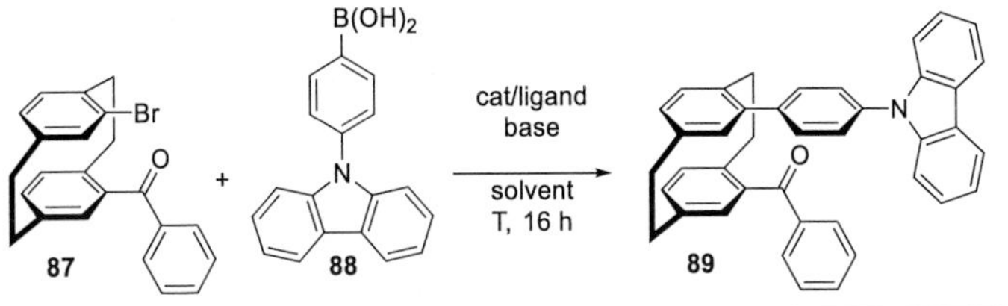

Entry	Catalyst/Ligand	Base	solvent	Temperature [°C]	Isolated product yield [%]
1	Pd(PPh$_3$)$_4$ 6 mol%	K$_3$PO$_4$ (1.5 equiv.)	dioxane/H$_2$O	110	–
2	Pd(PPh$_3$)$_4$ 6 mol%	K$_3$PO$_4$ (4.0 equiv.)	DMF	110	–
3	Pd(PPh$_3$)$_4$ 6 mol%	K$_3$PO$_4$ (4.0 equiv.)	DMF	50	–
4	Pd(OAc)$_2$ 5 mol% RuPhos 10 mol%	K$_3$PO$_4$ (2.0 equiv.)	toluene/H$_2$O	75	5
5	Pd(OAc)$_2$ 10 mol% RuPhos 20 mol%	K$_3$PO$_4$ (2.0 equiv.)	toluene/H$_2$O	75	57

Using the optimized Suzuki-Miyaura cross-coupling conditions, the diphenylamino derivative **91** was synthesized with the analogous boronic acid **90** (Scheme 34). Interestingly, the standard catalyst loading of 5 mol% Pd(OAc)$_2$ / 10 mol% RuPhos (compare Table 1, entry 4) was sufficient to obtain the second emitter **91** in 65% yield as a blue luminescent compound.

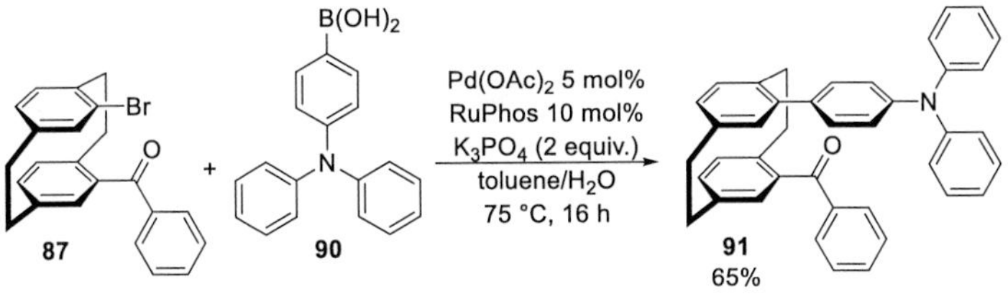

Scheme 34. Suzuki-Miyaura cross-coupling to emitter **91**.

With these two pseudo-*geminal* (*cis*) emitters in hand, the pseudo-*para* (*trans*) isomers were synthesized next.

3.1.2.2   Synthesis of pseudo-*para* Emitters

In the first step, a twofold unselective electrophilic bromination of [2.2]paracyclophane (**1**) was conducted as seen in Scheme 35. As the second bromination proceeded at the second ring with little selectivity,[34] a mixture of numerous dibrominated isomers was obtained, from which the desired pseudo-*para* dibromide **22** could be easily isolated by fractional recrystallization, as this centrosymmetric compound did not show a dipole moment, leaving this molecule very insoluble. From recrystallization out of toluene, compound **22** was isolated in 28%.

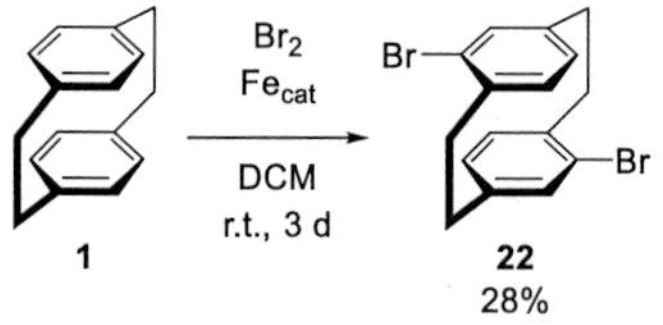

Scheme 35. Twofold electrophilic bromination of [2.2]paracyclophane (**1**) to access the pseudo-*para* dibromide **22**.

In the next step, one bromide was lithiated by addition of *n*-butyllithium and upon completed transmetallation, which can be monitored by color change to an intense yellow solution, benzoyl chloride was added in one portion to electrophilically quench the reactive lithiated intermediate **92**. The desired bromo benzoyl derivative **93** with a pseudo-*para* substitution pattern was obtained in 68% yield.

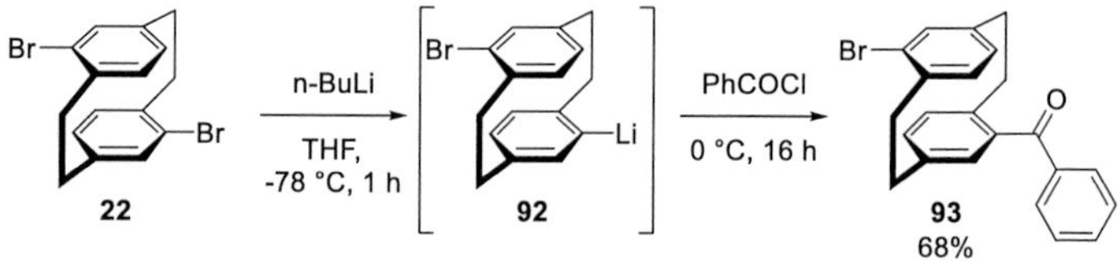

Scheme 36. Synthesis of pseudo-*para* bromo-benzoyl intermediate **93**.

From the key intermediate **93**, Suzuki-Miyaura cross-coupling protocols were screened in Table 2. The initial protocol was taken from own results (see chapter 3.5)[44] using $Pd(PPh_3)_2Cl_2$ and K_2CO_3 as base at 100 °C (entry 1). Again, only dehalogenated product was detected. Reducing the temperature to 75 °C and a switch to Cs_2CO_3 as base fortunately gave the desired emitter in 76% yield (entry 2). The isolated compound showed to be nonluminescent again.

Table 2. Screening of Suzuki-Miyaura cross-coupling conditions to emitter **94**.

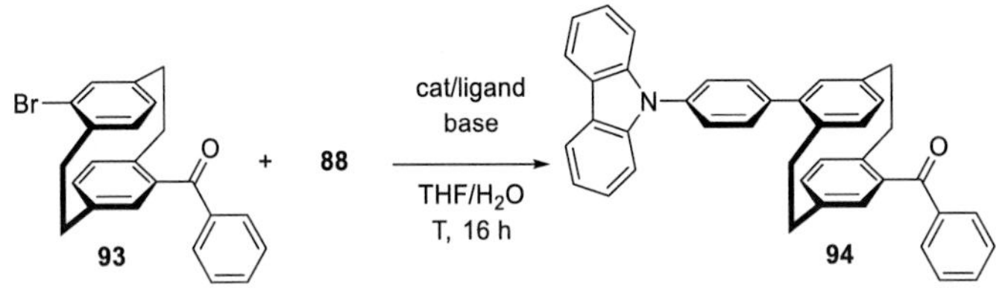

Entry	Catalyst/Ligand	Base	Temperature [°C]	Isolated product yield [%]
1	Pd(PPh$_3$)Cl$_2$ 10 mol%	K$_2$CO$_3$ (3.00 equiv.)	100	–
2	Pd(PPh$_3$)Cl$_2$ 10 mol%	Cs$_2$CO$_3$ (2.00 equiv.)	75	76

Again the cross-coupling protocol which turned out to be successful with the carbazolyl boronic acid **88** was also used for the analogous 4-(*N,N*-diphenylamino) phenyl boronic acid **90** (Scheme 37), yielding the desired emitter **95** in 65% isolated yield as a blue luminescent compound.

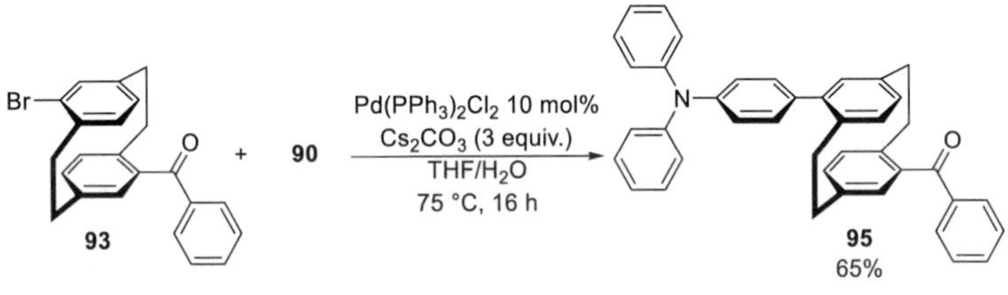

Scheme 37. Synthesis of the pseudo-*para* (trans) diphenylamino emitter **95**.

With the four potential emitters in hand, characterization was performed next. For this a structure based (e.g. *cis/trans*-Donor-Acceptor-Core) nomenclature was introduced.

3.1.2.3   Characterization

Calculations and spectroscopic analyses were performed *via* collaboration with the Zysman-Colman group in St. Andrews, Scotland. All experiments in this chapter were conducted by Nidhi Sharma. The nomenclature of the synthesized emitters were changed to a structure-based system with pseudo-*geminal* being *cis*, pseudo-*para* being *trans*, the diphenylamino and carbazolyl groups are abbreviated as DPA, and Cz, respectively.

DFT Calculations

DFT calculations were performed with the Gaussian 09 revision D.018 suite.[130] Initially the geometries in the ground state in gas phase were optimized employing the PBE0[131] functional with the standard Pople 6-31G(d,p) basis set.[132] Time-dependent DFT calculations were performed using the Tamm–

Dancoff approximation (TDA).[133] The graphical results are given in Figure 24, Figure 25 and Figure 26. A summary of relevant values is compiled in Table 10.

The first thing to notice is that the LUMOs in every emitter are concentrated on a benzophenone moiety, occupying only one of the two [2.2]paracyclophane decks. The same observation is made for the Cz and DPA donors, which are expanded onto one of the [2.2]paracyclophane decks as well, giving *syn* (*cis*) and *anti* (*trans*) parallel orientation of the respective frontier molecular orbitals (FMO)s.

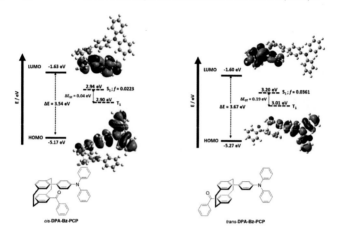

Figure 15. DFT calculations of the diphenylamine donor-based isomers **91** (*cis*-**DPA-Bz-PCP**) and **95** (*trans*-**DPA-Bz-PCP**). Results and images provided by Dr. Sharma.[134]

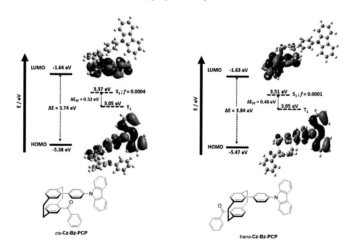

Figure 16. DFT calculations of the diphenylamine donor-based isomers **89** (*cis*-**Cz-Bz-PCP**) and **94** (*trans*-**DPA-Bz-PCP**). Results and images provided by Dr. Sharma.[134]

When the energy values (Table 3) of the four potential TADF emitter ware compared, then an increase of the $\Delta E_{\text{LUMO-HOMO}}$ gaps observed from the respective *cis* to *trans* isomers, with both DPA (3.54 eV for *cis* and 3.67 eV for *trans*, entries 1 and 2) and Cz emitters (3.74 eV for *cis* and 3.84 eV for *trans*, entries 3 and 4) with expected emission in the blue spectral range. In contrast, the ΔE_{ST} values essential for predicting the TADF efficiency, predict that only the DPA emitters (entry 1 and 2) can be considered TADF.

Table 3. Summary of DFT results for the benzoyl-acceptor based emitters.

Entry	Emitter	$\Delta E_{\text{LUMO-HOMO}}$ [eV]	ΔE_{ST} [eV]	Observation
1	*cis*-**DPA-Bz-PCP**	3.54	0.04	luminescent
2	*trans*-**DPA-Bz-PCP**	3.67	0.19	luminescent
3	*cis*-**Cz-Bz-PCP**	3.74	0.32	nonluminescent
4	*trans*-**Cz-Bz-PCP**	3.84	0.46	nonluminescent

Spectroscopic Results

Accordingly, only the luminescence performance of the DPA emitters (*cis*-**DPA-Bz-PCP**, **91** and *trans*-**DPA-Bz-PCP**, **95**) was evaluated, as these compounds were the only ones luminescent.

Photoluminescence and absorbance spectra of emitters *cis*-**DPA-Bz-PCP** (**91**) as well as *trans*-**DPA-Bz-PCP** (**95**) were recorded in toluene (Figure 17, left). Both emitters show absorbance at 311 nm and 312 nm, respectively. The emission peaks are located at 480 nm and 465 nm, respectively, complying with the predicted higher FMO gap for the *trans*-isomers. The emission shoulder at 410 nm and 404 nm, respectively, was assigned to the "phane state" formed between two benzene decks of the [2.2]paracyclophane scaffold.[135-136] The photoluminescence quantum yields (PLQY, Φ_{PL}) in degassed toluene solution were 45% and 60% for the *cis*-**DPA-Bz-PCP** (**91**) and *trans*-**DPA-Bz-PCP** (**95**) emitters. These values dropped to 30% and 42%, respectively, upon exposure to air. Photoluminescence emission from the solid state was determined for the unpolar hosts PMMA (as 10 wt% dopant) and mCP (as 15 wt% dopant) as shown in Figure 17, on the right side. While PMMA gave low PLQYs for both emitters with emission wavelengths of 485 nm (5%), and 470 nm (7.5%) for the *cis*- and *trans*-isomers of DPA, employing the higher triplet energy host mCP only gave mediocre improvement of PLQY with emission wavelengths of 480 nm (12%), and 465 nm (15%) for *cis*-**DPA-Bz-PCP** (**91**) and *trans*-**DPA-Bz-PCP** (**95**) emitters, respectively.

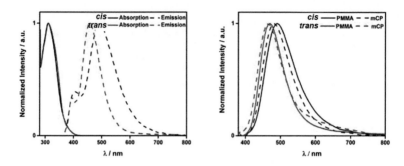

Figure 17. Photoluminescence spectra of the DPA emitters *cis*-**DPA-Bz-PCP** (**91**) and *trans*-**DPA-Bz-PCP** (**95**) in toluene (left) and as dopants in PMMA and mCP host films (right), λ_{exc} = 360 nm. Results and images provided by Dr. Sharma.[134]

Due to these strongly reduced PLQY values, other common host materials were tested (Table 4), such as DPEPO (entry 2), which gave PLQY values of 5% and 4.5% for *cis*-**DPA-Bz-PCP** (**91**) and *trans*-**DPA-Bz-PCP** (**95**), respectively, PPT (entry 3) with 7% and 8.3% and CzSi (entry 4) with 8.6% and 9.3%. Ultimately, mCP (entry 5) proved to be the optimal host of this screening with 12 and 15%, respectively.

Table 4. Screening PLQY for the benzoyl acceptor-based emitters in various host materials. Thin films were prepared by vacuum deposition and values were determined using an integrating sphere (λ_{exc} = 360 nm). Results and images provided by Dr. Sharma.[134]

Entry	Host Material	Φ_{PL} [%]	
		cis-**DPA-Bz-PCP** (**91**)	*trans*-**DPA-Bz-PCP** (**95**)
1	10 wt% PMMA	5	7.5
2	5 wt % DPEPO	5.0	4.5
3	7 wt% PPT	7.0	8.3
4	10 wt% CzSi	8.6	9.3
5	15 wt% mCP	12.2	15.0

In order to experimentally determine the TADF character, ΔE_{ST} was determined from the onset of the fluorescence (promt) and phosphorescence (delayed) spectra (Figure 18), measured in 15 wt% mCP films. The determined ΔE_{ST} value of 0.17 eV for *cis*-**DPA-Bz-PCP** (**91**) is in accordance to the calculated value of 0.19 eV, while the experimentally determined ΔE_{ST} of 0.28 eV for *trans*-**DPA-Bz-PCP** (**95**) is significantly smaller than the predicted 0.46 eV.

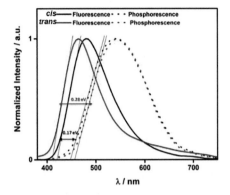

Figure 18. Prompt and delayed spectra at 77 K of *cis*-**DPA-Bz-PCP** (**91**) and *trans*-**DPA-Bz-PCP** (**95**) (λ_{exc} = 378 nm).in 15 wt% mCP film. Results and images provided by Dr. Sharma.[134]

Transient photoluminescence measurements in 15 wt% doped mCP films at various temperatures (Figure 19) revealed biexponential decay kinetics. Both isomers exhibited a prompt lifetime, τ_P, of 17 ns (*cis*-**DPA-Bz-PCP** (**91**)) and 7.4 ns (*trans*-**DPA-Bz-PCP** (**95**)) and a very short delayed lifetime, τ_D, of 1.8 µs and 3.6 µs, respectively. Such short τ_D values are characteristics of an efficient reversed intersystem crossing mechanism. The increased intensity of delayed emission with rising temperature is a direct evidence of the TADF properties of the two synthesized emitters.

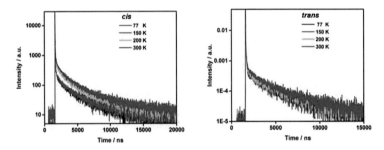

Figure 19. Photoluminescence decay curves of *cis*-**DPA-Bz-PCP** (**91**) and *trans*-**DPA-Bz-PCP** (**95**) 15 wt% mCP thin films; (λ_{exc} = 378 nm). Results and images provided by Dr. Sharma.[134]

The photophysical results discussed in this chapter are summarized in Table 5.

Table 5. Summary of spectroscopical properties of the benzoyl acceptor-based luminescent emitters *cis*-**DPA-Bz-PCP** (**91**) and *trans*-**DPA-Bz-PCP** (**95**).

Entry	Property	*cis*-**DPA-Bz-PCP** (**91**)	*trans*-**DPA-Bz-PCP** (**95**)
1	$\lambda_{abs\ (toluene)}$ [a] [nm]	311	312
2	$\lambda_{PL\ (toluene)}$ [a] [nm]	480	465
3	$\Phi_{PL\ (toluene)}$ [b] [%]	45	60
4	$\lambda_{PL\ (mCP\ thin\ film)}$ [c] [nm]	480	465
5	$\Phi_{PL\ (mCP\ thin\ film)}$ [c] [%]	12	15
6	τ_P [d] [ns]	17	7.4
7	τ_d [d] [μs]	1.8	3.6

[a] In degassed toluene at 298 K. [b] 0.5 M quinine sulfate in H_2SO_4 (aq) was used as the reference (Φ_{PL}: 54.6%, λ_{exc} : 360 nm).[137] Values quoted are in degassed solutions, which were prepared by three freeze-pump-thaw cycles. Values in parentheses are for aerated solutions, which were prepared by bubbling air for 10 min. [c] Thin films were prepared by vacuum depositing 15 wt% doped samples in mCP and values were determined using an integrating sphere (λ_{exc} : 360 nm); degassing was done by N_2 purge. [d] Values in parentheses are the pre-exponential weighting factors, determined in 15 wt% mCP doped films (λ_{exc} = 378 nm).

In conclusion, the first examples of TADF emitters incorporating a [2.2]paracyclophane core were synthesized exploiting the through-space electronic communication concept between donor and acceptor groups on adjoining benzene decks. Both the *cis*- and *trans*-isomers exhibited blue TADF emission and short delayed lifetimes in the range of 1.8–3.6 μs. Unfortunately, these compounds possess low photoluminescence quantum yields, especially in the solid state. No OLEDs were fabricated due to the poor luminance of the emitters. The results of a competing group published after this work, reporting diphenylamino donor and dimesityl acceptor units (see chapter 1.3.2.1 in the introduction) showed improved photoluminescence quantum yield of 53% for the pseudo-*geminal* isomer and 33% for the pseudo-*meta* isomer. This was accompanied by extraordinarily short delayed emission lifetimes (τ_d) of 0.38 μs and 0.22 μs, respectively, which may be considered not TADF, but simple fluorescence.[110]

3.1.3 Nitrile Acceptor for TADF

3.1.3.1 Synthesis of pseudo-*geminal* Emitters

The next acceptor group to be synthesized in pseudo-*geminal* and pseudo-*para* orientations was nitrile. In contrast to the benzoyl, the nitrile is not reported as a transannular directing group. Additionally, [2.2]paracyclophane nitriles are commonly synthesized *via* Rosenmund von Braun reaction using bromo[2.2]paracyclophane as the starting material and treating it with an excess of copper(I)cyanide in highly boiling solvents to get moderate yields of the product,[126, 138] at which thermal isomerization can occur.[9] Furthermore, the bromide is also not a transannular directing group. Therefore, a suitable carbonyl functional group had to be employed in the synthetic route as an intermediate that would direct the electrophilic bromination to the pseudo-*geminal* position and could be subsequently converted to the nitrile (see Scheme 29 for the retrosynthetic analysis). Parts of the initial synthetic approach were supervised previous work,[139] but all molecules reported here have been also prepared by myself. To do that, a procedure converting synthetically very versatile aldehydes to nitriles *via* dehydration of aldoxime intermediates was tried on formyl[2.2]paracyclophane **5** as a model system, which was synthesized *via* a standard Rieche protocol to obtain the desired compound in 90% yield (Scheme 38).[48]

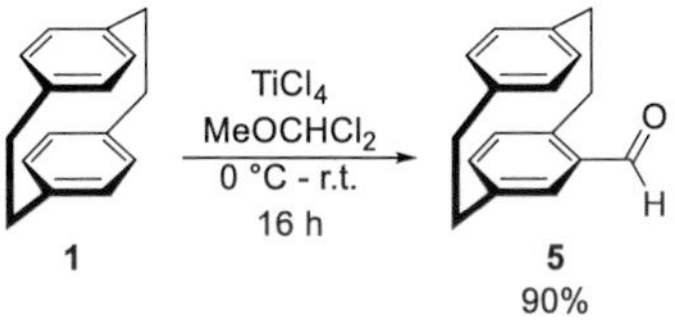

Scheme 38. Rieche formylation of 4-formyl[2.2]paracyclophane (**5**).

The aldehyde **5** was screened with different amounts of hydroxylammonium chloride, reaction time and temperature (Table 6). Upon addition of equimolar amounts (entry 1), no product was detected. An increase of equivalents to a three-fold excess (entry 2), led to the nitrile **97** in 22% yield. Further increase of the excess to four-fold and doubling the reaction time (entry 3), pushed the yield to 55%. Raising the temperature to 100 °C (entry 4), did not significantly improve the yield (57%), but a decomposition of DMSO was detected as a foul smell. Therefore, it was found that the protocol of the aldehyde conversion to the nitrile was possible on the [2.2]paracyclophane core, and a synthetic path to a pseudo-*geminal* substituted key intermediate could be conducted *via* a bromo aldehyde intermediate. No further optimization of this reaction was conducted, but could be reached by addition of acid or a hydrophilic reagent to promote the dehydration of the aldoxime intermediate **96**.

Table 6. Screening of equivalents of hydroxylammonium chloride to obtain nitrile **97**.

Entry	HONH$_3^+$Cl$^-$ equivalents	Temperature [°C]	Time [h]	Isolated product yield [%]
1	1.0	90	1	–
2	3.3	90	1	22
3	4.0	90	2	55
4	4.0	100	2	57

Ideally, a direct bromination of **97** to **103** would be favorable. As no reports of bromination of the nitrile **5** were found and therefore also no directing efficiency is reported, a screening of the direct bromination was conducted (Table 7). As a diminished (if any) directing effect was expected, the goal was also to see if a resulting regioisomer mixture could be separated. Upon standard conditions (entry 1), an immediate coloration to violet / black was observed, yet no product mass was detected in GC-MS. An increase of the reaction temperature to refluxing DCM showed identical coloration, with only traces of product mass in GC-MS. Due to the suspicious coloration of the reaction mixture, the direct route was not further investigated.

Table 7. Screening of the direct bromination of nitrile **97**.

Entry	Conditions	Yield [%]
1	Fe$_{cat}$, Br$_2$ (1 equiv.) DCM, r.t. 16 h	–
2	Fe$_{cat}$, Br$_2$ (1 equiv.) DCM, 40 °C, 16 h	Traces in GC-MS

As the next best route, a direct bromination of the aldehyde **5** would yield the suitable intermediate **99**, but in contrast to other carbonyl compounds such as ketones, esters and amides, again no transannular directing bromination of **5** was reported in literature. Upon a test screening (Table 8), no product could be isolated, essentially giving the same results as observed for the nitrile **97** under identical conditions. Upon standard bromination conditions (entry 1), no product could be detected, but upon addition of bromine, the solution turned dark purple. Upon a slight increase of temperature, traces of the product mass were detected in GC-MS. Upon switching the bromination conditions to acetic acid as a solvent with a higher boiling temperature, also only traces of the product mass could be observed, again accompanied by a suspicious dark purple coloration of the reaction mixture. Due to this, no further screening was performed, but another multi-step route was conducted.

Table 8. Screening of the direct bromination of aldehyde **5**.

| | | **5** | **99** | |

Entry	Conditions	Yield [%]
1	Fe$_{cat}$, Br$_2$ (1 equiv.)	–
	DCM, r.t. 16 h	
2	Fe$_{cat}$, Br$_2$ (1 equiv.)	Traces in GC-MS
	DCM, 40 °C, 16 h	
3	Br$_2$ (1 equiv.)	Traces in GC-MS
	HOAc. 4 h	

To circumvent this, a route of Rowlands *et al.* was conducted to get intermediate **99**.[140] According to this, a synthetic route to obtain the pseudo-*geminal* substituted bromo nitrile intermediate **103** was planned as seen in Scheme 39. Carbomethoxy ester **100** was identified to be a convenient substrate for the transannular directed pseudo-*geminal* bromination, as it can be obtained from the [2.2]paracyclophane (**1**) in three steps *via* lithiation of the brominated derivative with subsequent quenching with CO$_2$ to the respective free acid. As this free acid is known to have poor solubility, an esterification is performed to obtain **100**. This intermediate is known to direct electrophilic bromination to the pseudo-*geminal* position, yielding **101**.[141] Finally, the ester has to be reduced to the aldehyde function to obtain **99**, which can be converted to the final nitrile acceptor intermediate with a pseudo-*geminal* bromine *via* the protocol depicted in Table 6.

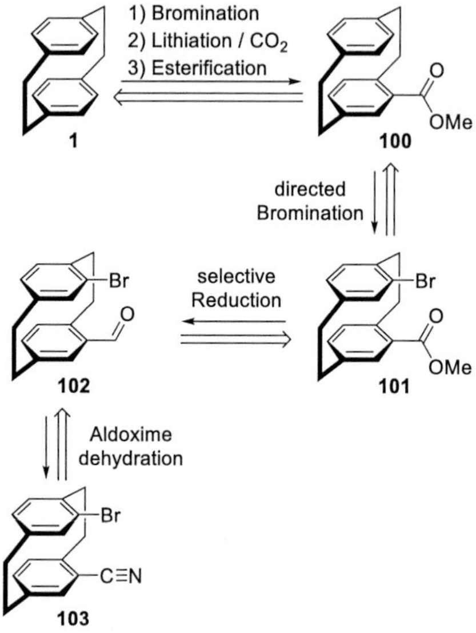

Scheme 39. Retrosynthetic analysis towards bromo aldehyde **99**[140] and subsequent conversion to the bromo nitrile acceptor intermediate **103** in the desired pseudo-*geminal* orientation.

First, the literature-known Fe-catalyzed electrophilic bromination (Scheme 40) of [2.2]paracyclophane (**1**)[25] on a 15 g scale is described, yielding the bromo derivative **17** in an excellent yield of 98%.

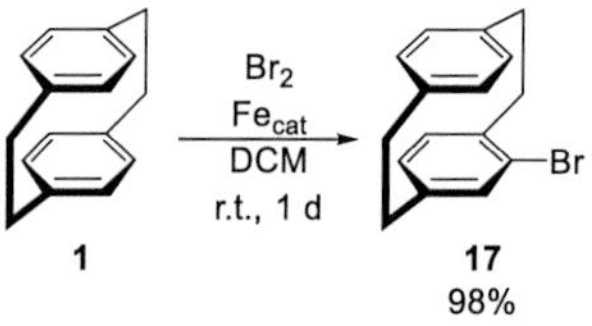

Scheme 40. Synthesis of bromo-[2.2]paracyclophane **17**.

Subsequently, lithiation of the bromo group, followed by electrophilic quenching with CO_2 gave the free acid **105** in a moderate yield of 65% (Scheme 41).

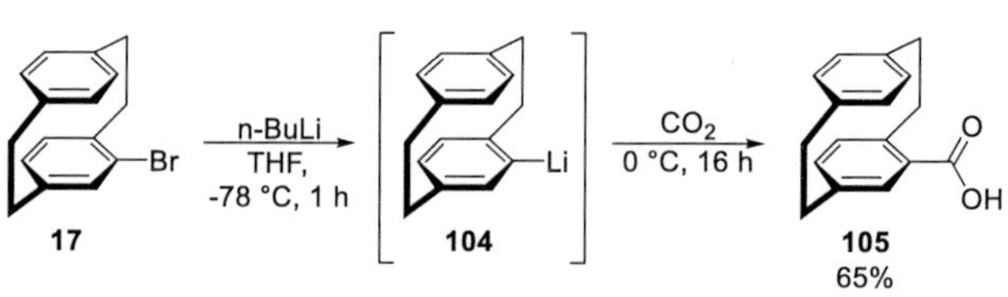

Scheme 41. Synthesis of 4-carboxy[2.2]paracyclophane (**105**).

To access the carbomethoxy intermediate **100** needed for transannular directive electrophilic bromination, the free acid **105** was refluxed in methanol under acidic conditions to promote the esterification, which gave the product in an excellent yield of 92%.

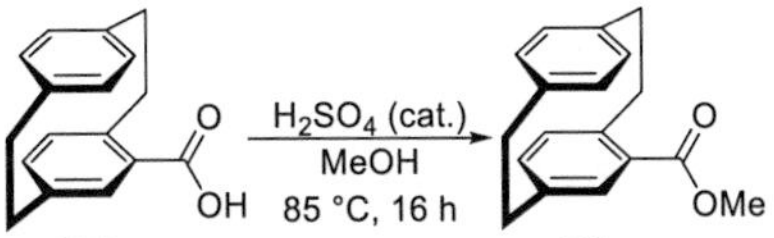

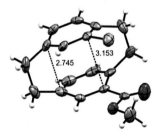

105   H$_2$SO$_4$ (cat.)   **100**
            MeOH   92%
            85 °C, 16 h

Scheme 42. Esterification of the free acid **105**.

The iron-catalyzed electrophilic bromination of ester **100** was conducted following the standard procedure of the Bräse group (Scheme 43).[142] The desired pseudo-*geminal* substituted intermediate **101** was obtained in moderate 49% yield. The resulting yield is comparable to the pseudo-*geminal* bromination of the benzoyl intermediate **87** (see Scheme 33) which gave 46% yield. Compared to the recent literature, an increase of yield can be achieved by reduced reaction time and dry solvent.[140]

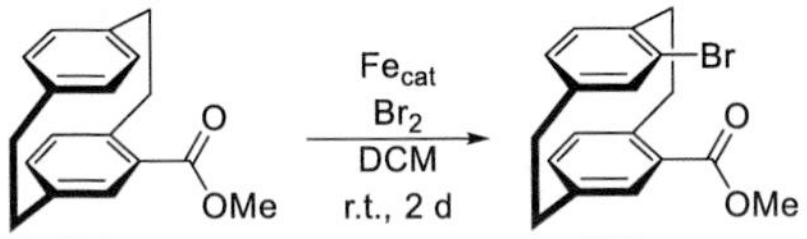

100   Fe$_{cat}$ **101**
           Br$_2$ 49%
           DCM
           r.t., 2 d

Scheme 43. Transannular-directed electrophilic bromination of ester **100** to the pseudo-*geminal* substituted bromo ester **101**, which could be analyzed by single crystal analysis, verifying the pseudo-*geminal* orientation of the substituents (Figure 20).

Figure 20. Molecular structure of pseudo-*geminal* substituted **101** drawn at 50% probability level.

With the intermediate **101** in hand, different strategies of reduction towards the aldehyde were conducted. The direct selective reduction (Scheme 44) using DIBAL-H showed no reduction. As the reduction of esters with DIBAL-H is known to be a capricious procedure[143] and since DIBAL-H is a bulky reagent, no further investigation of this direct reduction route was carried out, but another approach from the fully reduced alcohol was performed.

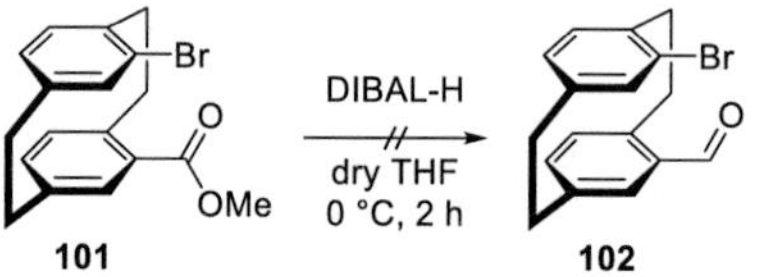

Scheme 44. Unsuccessful direct reduction to the aldehyde **99**.

To get the fully reduced alcohol **106**, LiAlH$_4$ was used (Scheme 45, top).[140] As expected from the reliably reduction using LiAlH$_4$, an excellent yield of 95% was obtained. Interestingly, in contrast to the parent alcohol without the bromine atom the alcohol **106** shows a $^2J_{HH}$ coupling of the two diastereotopic hydrogen atoms in the ^{1}H NMR (Scheme 45, bottom), which is an example of the steric strain present upon pseudo-*geminal* difunctionalization.

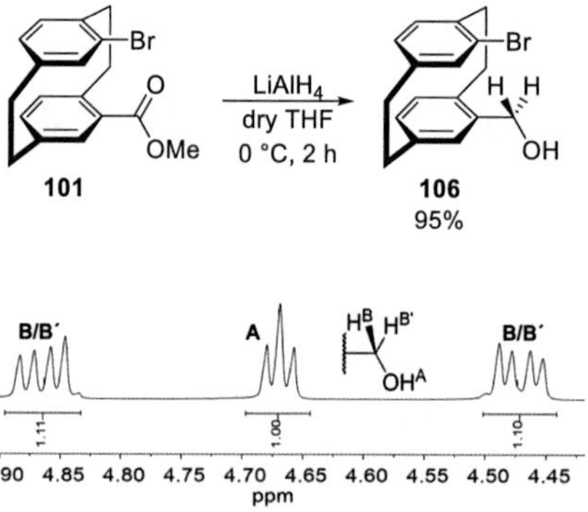

Scheme 45. Full reduction of ester **101** to alcohol **106** with an extract of the ^{1}H NMR spectrum in DMSO, showing the magnetically inequivalent methylene hydrogen atoms.

Next, aldehyde **99** was obtained by Swern oxidation of the alcohol **106** (Scheme 46) in 63% yield.

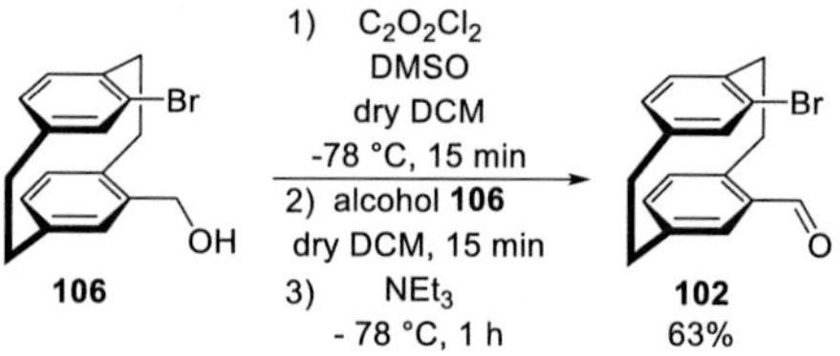

Scheme 46. Swern oxidation to aldehyde **102**.

Finally, the conversion of the aldehyde **99**, which was synthesized over six steps, could be performed according to the optimized procedure from Table 6 in a moderate yield of 54%. Compared to the 57% yield obtained for the less encumbered model system **97**, no significant drop of yield was observed.

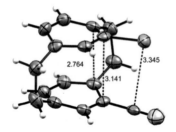

Scheme 47. Synthesis of the pseudo-*geminal* disubstituted bromo nitrile key intermediate **103**.

Furthermore, the absolute configuration was verified by single crystal analysis (Figure 21). Compared to the respective pseudo-*geminal* bromo benzoyl compound **87**, the distances of the bridgehead atoms are identical with 2.76 Å, while the deck distance at the substituent bearing carbon atoms are slightly smaller for the nitrile **103** with 3.14 Å *versus* 3.17 A for the benzoyl derivative **87**. The distance of the bromine atom to the nitrile carbon is 3.35 Å, which is slightly bigger than for the deck distance, but still considerably close, in comparison to the identical layer distances in graphite.[19] The distances are nearly identical to the ester precursor **101** (Figure 20).

Figure 21. Molecular structure of pseudo-*geminal* substituted **103** drawn at 50% probability level.

Finally, Suzuki-Miyaura cross-coupling could be performed with the pseudo-*geminal* key intermediate **103** as presented in Table 9 using the 4-(*N,N*-diphenylamino) phenyl boronic acid **90** with an excess of three equivalents as the diphenylamino donor unit proved to be a more promising TADF donor unit. The initial conditions, which were successful for the pseudo-*geminal* bromo benzoyl intermediate **87**, gave with only 4% a minimal amount of isolated yield (entry 1). Upon increasing the reaction temperature to 85 °C and doubling the amount of the catalyst system (entry 2), no product was isolated. Excluding water from the reaction and increasing the reaction time to three days (entry 3) still did not help. A change of the catalyst to Pd(PPh₃)₄ in dry toluene at 75 °C (entry 4) gave an improved yield of 22%. This could be further enhanced by an increase of the catalyst loading to 2 × 30 mol%, heating to 85 °C and a prolonged reaction time of 6 days, yielding moderate 63% (entry 5). A reduction of the catalyst loading to 30 mol% accompanied by a reaction time of 10 days (entry 6) resulted in a slightly diminished yield of 56%. As enough material was isolated during the screening and since the moderate yields of above 50% were considered sufficient in order to obtain the emitter, no further optimization was conducted. The isolated compound showed to be deep-blue luminescent.

Table 9. Screening of Suzuki-Miyaura cross-coupling conditions to pseudo-*geminal* carbazolyl nitrile emitter **107**.

Entry	Catalyst/Ligand/Base	Solvent	Temperature [°C]	Time	Isolated product yield [%]
1	Pd(OAc)$_2$ (5 mol%) RuPhos (10 mol%) K$_3$PO$_4$ (2 equiv.)	Toluene/H$_2$O	75	16 h	–
2	Pd(OAc)$_2$ (10 mol%) RuPhos (20 mol%) K$_3$PO$_4$ (2 equiv.)	Toluene/H$_2$O	85	16 h	–
3	Pd(OAc)$_2$ (5 mol%) RuPhos (10 mol%) K$_3$PO$_4$ (2 equiv.)	Toluene	85	3 d	–
4	Pd(PPh$_3$)$_4$ (10 mol%) K$_3$PO$_4$ (2.5 equiv.)	Toluene	75	3 d	22
5	Pd(PPh$_3$)$_4$ (30 mol%) K$_3$PO$_4$ (2.5 equiv.)	Toluene	85	6 d	63[a]
6	Pd(PPh$_3$)$_4$ (30 mol%) K$_3$PO$_4$ (2.5 equiv.)	Toluene	85	10 d	56

[a]: A second aliquot of catalyst, base and boronic acid was added after 3 days.

The synthesis of the pseudo-geminal carbazolyl nitrile derivative **108** (Scheme 48) was conducted simultaneously to the screening of the optimization according to entry 4 with two equivalents of the analogous boronic acid **88** in a poor yield of 26%, which again gave enough material, as the isolated compound showed to be nonluminescent.

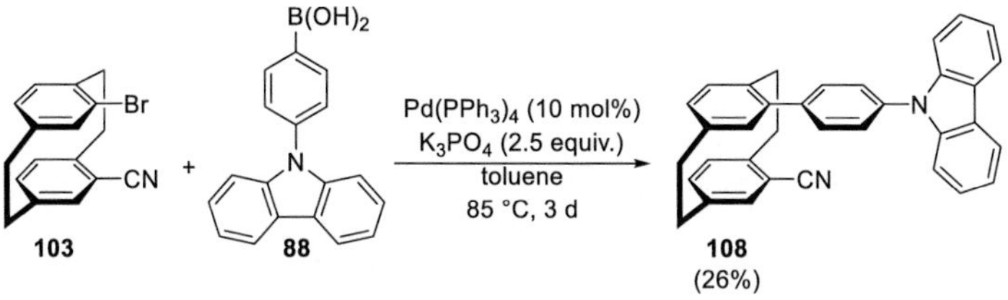

Scheme 48. Synthesis of the pseudo-*geminal* carbazolyl nitrile emitter **108**.

Additionally, a DMAC (9,9-dimethyl-9,10-dihydroacridine, or 9,9-dimethylacridan) donor variant **110** was also synthesized using 1.5 equivalents of the respective boronic acid **109** and 30 mol% of Pd(PPh₃)₄ at 85 °C for 3 days. Although the isolated yield was poor, no further optimization was performed, as the resulting emitter did not show any luminescence.

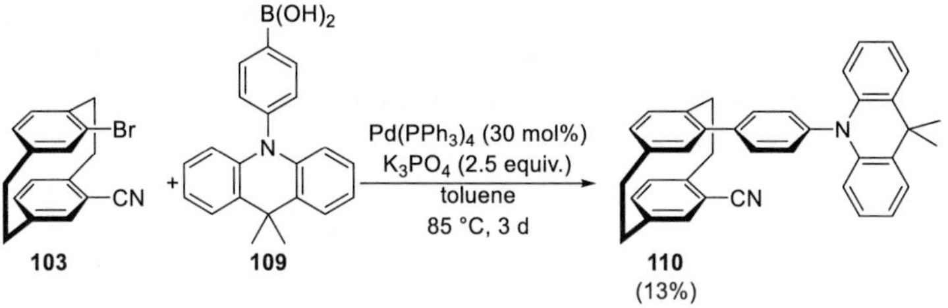

Scheme 49. Synthesis of the pseudo-*geminal* DMAC nitrile emitter **110**.

3.1.3.2   Synthesis of pseudo-*para* Emitters

In contrast, the pseudo-*para* nitrile-based emitters could be synthesized according to the retrosynthetic discussion in Scheme 30 with the aldoxime dehydration step optimized in Table 6. For this the pseudo-*para* dibromide **22** was lithiated and subsequently quenched with DMF to obtain the bromo aldehyde **112** in 69% yield (Scheme 50), which is similar to the yield achieved for the benzoylation (compare with Scheme 36).

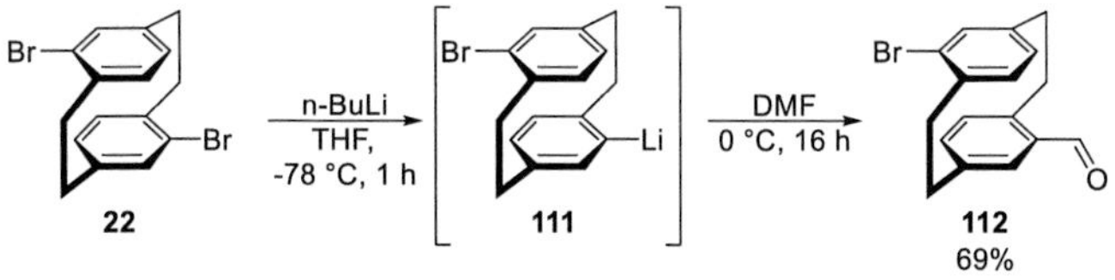

Scheme 50. Synthesis of pseudo-*para* bromo aldehyde **112** *via* monolithiation.

This intermediate was treated with hydroxylammonium chloride in DMSO at 100 °C to receive the pseudo-*para* bromo nitrile *via* an aldoxime dehydration in 60% yield (Scheme 51).

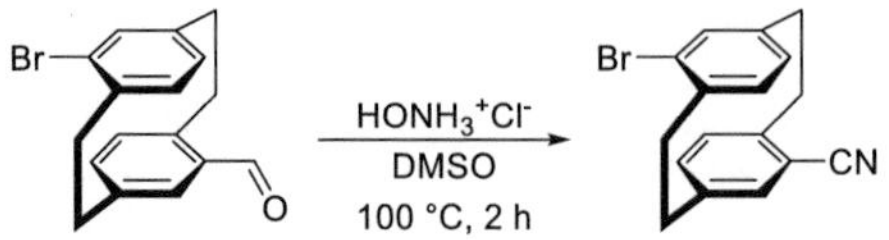

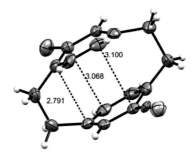

Scheme 51. Synthesis of pseudo-*para* bromo nitrile **113** *via* aldoxime dehydration.

In addition, the absolute configuration of **113** could be verified as pseudo-*para* by single crystal analysis (Figure 22). In comparison to the pseudo-*geminal* isomer **103**, the bridge-head distance of 2.79 Å is slightly increased from 2.76 Å. The deck distance measured as the distance between a functionalized carbon atom and the respective pseudo-*para* atom ranges between 3.10–3.07 Å, which is slightly smaller than the 3.14 Å distance measured for **103**. Due to the pseudo-*para* orientation of the two functional substituents, no additional steric effect is expected to occur during the cross-coupling to the final emitters.

Figure 22. Molecular structure of pseudo-*geminal* substituted **113** drawn at 50% probability level.

Gratifingly, in contrast to the needed screening of the Suzuki-Miyaura cross-coupling of pseudo-*geminal* oriented emitters shown before, the initial cross-coupling procedure using Pd(OAc)$_2$ / RuPhos gave the desired emitter **114** in excellent yield of 93% (Scheme 52) as a deep blue luminescent compound.

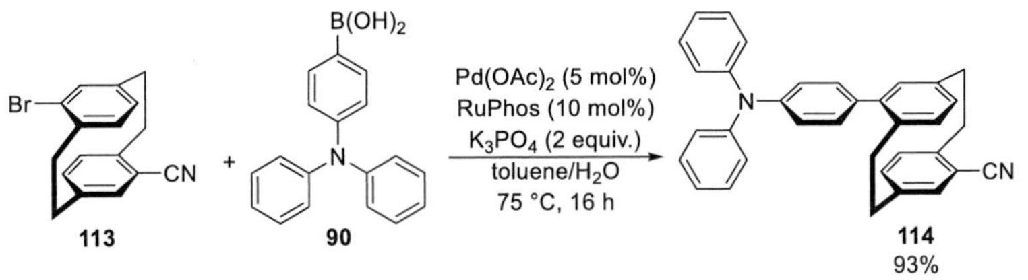

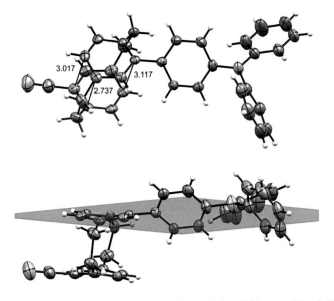

Scheme 52. Synthesis of the pseudo-*para* diphenylamino nitrile-based emitter **114**.

Single crystals of emitter **114** could be successfully isolated to determine its structure (Figure 23), showing that the bridge-head distance is slightly reduced to 2.73 Å, compared to the precursor **113**. Furthermore, it seems that the deck distance at the nitrile side is with 3.02 Å smaller than the phenyl-linked side with 3.12 Å, which is an inverted tilt of the planes in comparison to **103**. Interestingly, the triphenylamine donor group is also bent out of the upper [2.2]paracyclophane plane (Figure 23, bottom).

Figure 23. Molecular structure of pseudo-*geminal* substituted emitter **114** drawn at 50% probability level.

The carbazolyl emitter **115** was synthesized under identical conditions (Scheme 53) in a good yield of 75% as a nonluminescent compound,

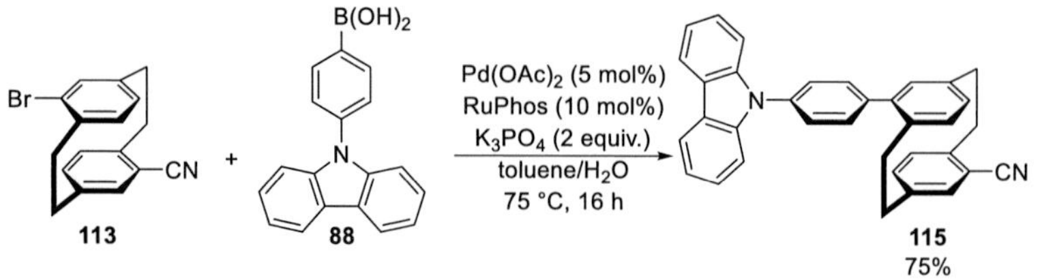

Scheme 53. Synthesis of the pseudo-*para* carbazolyl nitrile emitter **115**.

as well as the DMAC derivative **116**, which was isolated in 72% yield (Scheme 54), again as a nonluminescent compound.

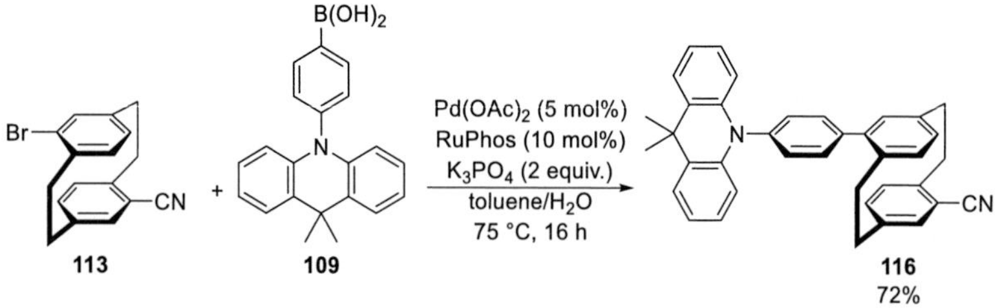

Scheme 54. Synthesis of the pseudo-*para* dimethylacridanyl nitrile emitter **116**.

3.1.3.3   Characterization

Calculations, and spectroscopic analyses were performed *via* collaboration with the Zysman-Colman group in St. Andrews, Scotland. All experiments in this chapter were conducted by Nidhi Sharma. The nomenclature of the synthesized emitters is changed to a structure-based system with pseudo-*geminal* being *cis*, pseudo-*para* being *trans*, the diphenylamino, carbazolyl and dimethylacridan groups are abbreviated as DPA, Cz, and DMAC, respectively.

DFT calculations

DFT calculations were performed with the Gaussian 09 revision D.018 suite.[130] Initially the geometries in the ground state in gas phase were optimized employing the PBE0[131] functional with the standard Pople 6-31G(d,p) basis set.[132] Time-dependent DFT calculations were carried out using the Tamm–Dancoff approximation (TDA).[133]

The graphical results are given in Figure 24, Figure 25 and Figure 26. A summary of relevant values is compiled in Table 10.

The first thing to notice is that in contrast to the LUMOs of the benzoyl acceptor-based emitters (see chapter 3.1.2.3), the LUMO is distributed over both [2.2]paracyclophane rings for all six emitters, essentially bridging the non-conjugated decks within the LUMO. Comparing the HOMOs, the weaker donor units DPA and Cz distribute the orbital partially onto the adjacent [2.2]paracyclophane deck, while the stronger DMAC donor contracts the HOMO away from the [2.2]paracyclophane.

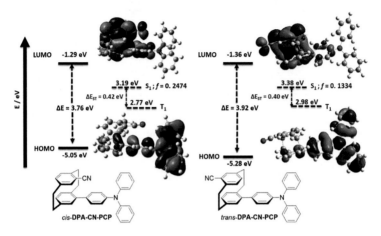

Figure 24. DFT calculations of the diphenylamino donor-based isomers **107** and **114**. Results and images provided by Dr. Sharma.[134]

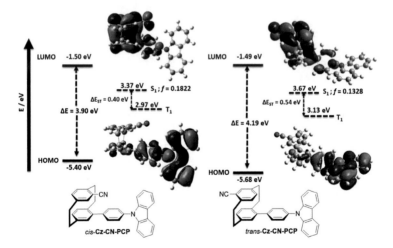

Figure 25. DFT calculations of the carbazolyl donor-based isomers **108** and **115**. Results and images provided by Dr. Sharma.[134]

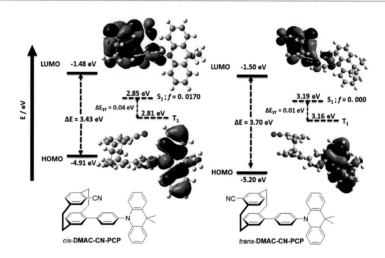

Figure 26. DFT calculations of the DMAC donor-based isomers **110** and **116**. Results and images provided by Dr. Sharma.[134]

When the energy values are compared (Table 10), an increase of the $\Delta E_{\text{LUMO-HOMO}}$ is observed from the respective *cis* to *trans* isomers, with both DPA (3.76 eV for *cis* and 3.92 eV for *trans*, entries 1 and 2) and Cz emitters (3.90 eV for *cis* and 4.19 eV *trans*, entries 3 and 4), all showing a huge energy gap at the edge of the visible spectrum. The increase of the FMO gap for the *trans*-isomers is also observed for the benzoyl acceptor-based emitter (see Table 3). With a $\Delta E_{\text{LUMO-HOMO}}$ of 4.19 eV, emitter *trans*-**Cz-CN-PCP** (entry 4) could be considered as a host material for OLEDs. In contrast, the DMAC donor unit reduces the HOMO-LUMO gap to 3.43 eV and 3.70 eV for the *cis*-and *trans*-isomers (entries 5 and 6), respectively, even leaving the blue emission spectral range. When comparing the ΔE_{ST} values which are essential for predicting the TADF efficiency, only the DMAC emitters (entry 5 and 6) can be considered TADF, while the four other emitters (entries 1–4) are far away from the desired minimized energy gap.

Table 10. Summary of DFT results for the nitrile-acceptor based emitters.

Entry	Emitter	$\Delta E_{\text{LUMO-HOMO}}$ [eV]	ΔE_{ST} [eV]	Observation
1	*cis*-**DPA-CN-PCP**	3.76	0.42	luminescent
2	*trans*-**DPA-CN-PCP**	3.92	0.40	luminescent
3	*cis*-**Cz-CN-PCP**	3.90	0.42	nonluminescent
4	*trans*-**Cz-CN-PCP**	4.19	0.54	nonluminescent
5	*cis*-**DMAC-CN-PCP**	3.34	0.04	nonluminescent
6	*trans*-**DMAC-CN-PCP**	3.70	0.01	nonluminescent

Spectroscopic Results

Only, the luminescence performance of the DPA emitters (*cis*-**DPA-CN-PCP**, **107** and *trans*-**DPA-CN-PCP**, **114**) was evaluated, as these were the only luminescent emitters.

The photoluminescence spectrum of emitter *cis*-**DPA-CN-PCP** (**107**) was recorded as a neat film and as a 10 wt% dopant in PMMA, showing blue emission of 425 nm and 426 nm, respectively. While the PLQY (Φ_{PL}) in the neat film was 26%, it could be increased to 52% in the unpolar host (Figure 27).

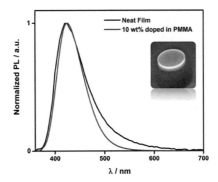

Figure 27. Photoluminescence spectra of *cis*-**DPA-CN-PCP** (**107**) as a neat film and as a dopant in PMMA with λ_{exc} = 360 nm. Results and images provided by Dr. Sharma.[134]

As expected from DFT calculations, the bigger $\Delta E_{LUMO-HOMO}$ gap for *trans*-**DPA-CN-PCP**, (**114**) resulted in a bluer emission with 410 nm for both the neat film and as a 15 wt% dopant in mCP (Figure 28). The photoluminescence quantum yield (Φ_{PL}) could be significantly increased from 7% in the neat film to 45% in the mCP host.

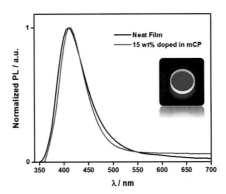

Figure 28. Photoluminescence spectra of *trans*-**DPA-CN-PCP** (**114**) as a neat film and as a dopant in mCP with λ_{exc} = 360 nm. Results and images provided by Dr. Sharma.[134]

An emission decay measurement (Figure 29) with a multiexponential fitting verified that due to the high ΔE_{ST} gap of 0.40 eV, no delayed fluorescence component was observed for *trans*-**DPA-CN-PCP** (**114**). Decay time τ was determined as 17 ns in neat film, and 598 ns in the mCP host. Although the decay time increased dramatically, it is too short to be considered TADF, as this process would yield τ_D values in the ms range.

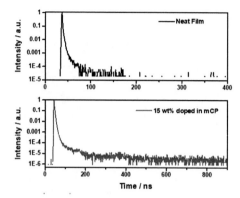

Figure 29. Fluorescence decay measurements of *trans*-**DPA-CN-PCP**, (**114**). Results and images provided by Dr. Sharma.[134]

In summary, the nitrile-based emitters suffer from poor luminescence in case of the aromatic amine donors, while the diphenylamine donor resulted in deep-blue emission, but only luminescent emitters, which do not possess any TADF properties. Despite this, the carbazolyl donor-based emitter exhibit a huge HOMO-LUMO gap, which make them suitable as host materials. Yet, in order to investigate this application, an upscale of the synthetic path is needed, as all emitters were isolated in small scale.

3.2 Carbazolophane Donor for TADF Turn-On

The synthesis of another fascinating [2.2]paracyclophane-based structure (**37**) was described by Bolm
et al. (Figure 30).[39] This potential donor is a planar chiral carbazole derivative with a
[2.2]paraphenylcyclophane moiety attached in the 1,4 position of the carbazole subunit.

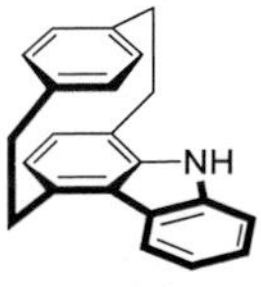

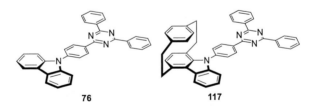

(rac)-**37**

Figure 30. [2]Paracyclo[2](1,4)carbazolophane (**37**).

Apart from chirality, the ethyl bridge adjacent to the aromatic amino group of the carbazole subunit
serves as a promising steric feature to introduce additional twist between the carbazole donor and a *N*-
coupled acceptor unit. Furthermore, the closely stacked second deck and the ethyl bridges both add
electron density to the donor unit and help distributing the HOMO away from the acceptor. As discussed
in section 1.3.2.2 in the introduction, the twist between a TADF donor unit and TADF acceptor unit is
a common design principle in TADF. The parent carbazolo triazine emitter **76** does not show TADF
properties, whereas a slight increase of twist around the *N*-donor–phenyl bond, or an increase of
dispersion volume for the donor turns on TADF properties of the structure. Therefore, the aim of this
project is the synthesis of triazine emitters **117** (Figure 31) and possible derivatives thereof with **37**
subunit in a racemic and if possible enantiopure manner. They will then be compared to the parent non-
TADF emitter **76**.

76   **117**

Figure 31. Targeted triphenyltriazine acceptor-based emitters.

3.2.1   Synthesis

In order to prepare the emitter **117**, a convergent strategy was conducted as seen in Scheme 55. The carbazolophane (**37**) donor and the 4-fluoro-triphenyltriazine (**119**) acceptor were synthesized separately and then these two subunits were combined *via* nucleophilic aromatic substitution promoted by the inherent electron withdrawing properties of **119**. As described by Bolm *et al.* the free carbazolophane **37** and numerous 4-derivatives can be synthesized by a multistep protocol starting from bromide **17**.[39] The bromide is first converted to the secondary [2.2]paracyclophanyl amine **36**, followed by an oxidative cyclization under acidic conditions to the key intermediate **37**. The 4-fluoro-triphenyltriazine (**119**) is readily obtained *via* Suzuki-Miyaura cross-coupling of the commercially available 2-chloro-4,6-diphenyltriazine (**121**) and 4-fluorophenylboronic acid (**120**). Parts of the initial synthetic approach were supervised in preliminary work,[144] but all molecules reported here have been also prepared by myself.

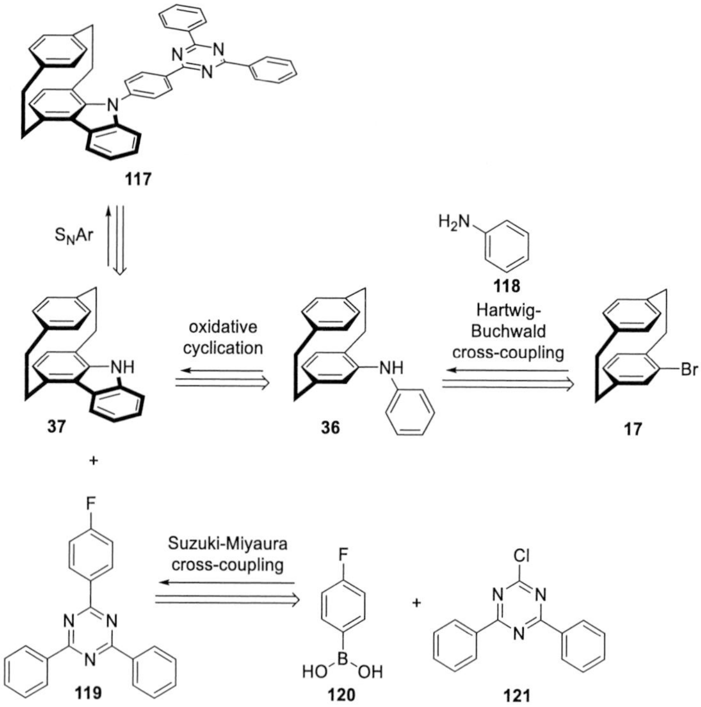

Scheme 55. Retrosynthetic analysis towards emitter **117**.

3.2.1.1   Synthesis of the Fluoro Triazine Acceptor

To synthesize the 4-fluoro triazine acceptor **119**, a literature known procedure by Lee *et al.* was followed,[145] with decreased water content and an excess of the boronic acid **120** as modifications (Scheme 56). The product was obtained in good yield on a multigram scale.

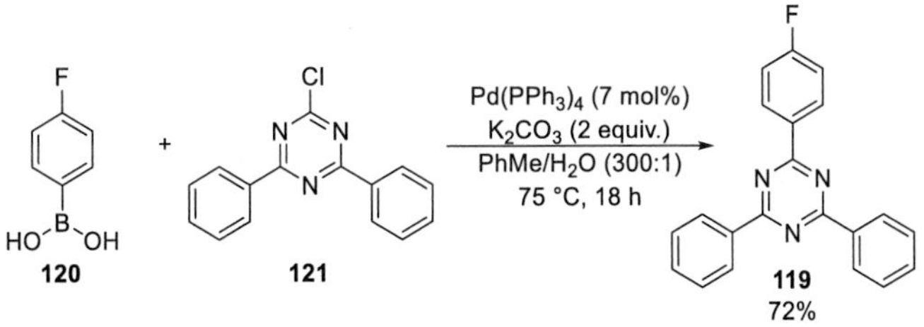

Scheme 56. Synthesis of fluoro triazine acceptor **119**.

3.2.1.2   Synthesis of 4-Carbazolyl-triphenyltriazine Model Emitter 76

First, the non-TADF model emitter **76** was synthesized by a nucleophilic aromatic substitution (Scheme 57) using DMSO and tripotassium phosphate as base, proven to be both convenient and efficient in our labs.

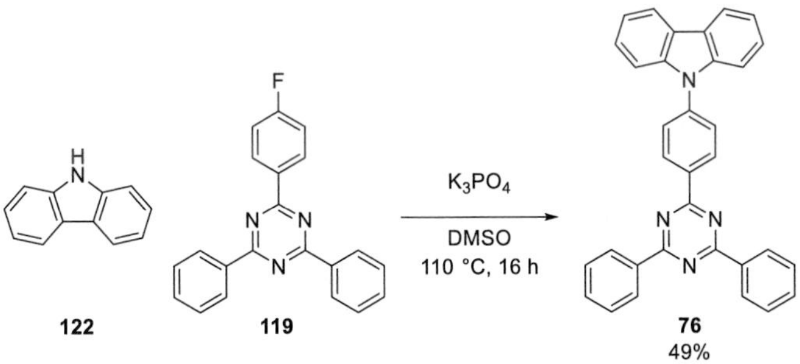

Scheme 57. Synthesis of the non-TADF model emitter **76**.

Even though the model emitter **76** is evaluated in literature, only one synthesis was described in a patent using NaH/DMF conditions for the chlorotriphenyltriazine as an alternative starting material with a similar yield of 56%,[146] showing that the 4-halo-triphenyltriazine group is a possible, but not very suitable electron-deficient system to promote nucleophilic aromatic substitution reactions. To further verify the structure, a single crystal analysis was performed as shown in Figure 32. The angle between the pyrrole ring plane of the carbazole donor unit and the bridging phenyl ring plane is 45.1°, while the angle between the triazine plane and the central phenyl bridge is only 5.5°. This gives a crystallographic benchmark in order to determine the twist between the donor unit and the bridging unit for the carbazolophane derivative.

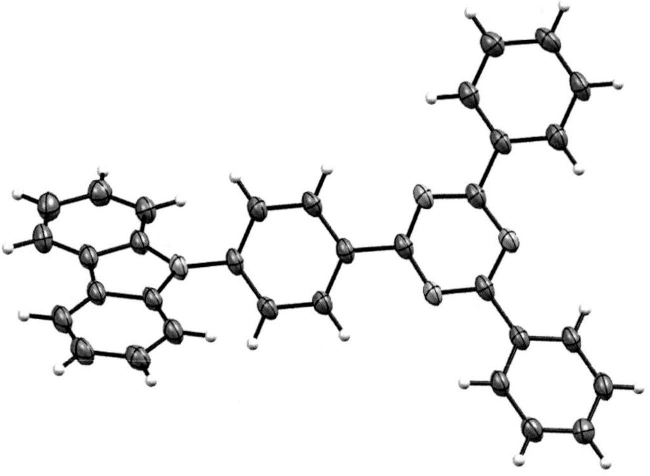

Figure 32. Molecular structure of emitter **76** drawn at 50% probability level.

3.2.1.3   Synthesis of Carbazolophanyl Triazine Emitter 117

In order to prepare carbazole **37**, a multistep synthesis was conducted starting with a Hartwig-Buchwald-cross-coupling of the previously prepared racemic bromide **17** with aniline (**118**) to yield the secondary amine **36** in 82% (Scheme 58).[39]

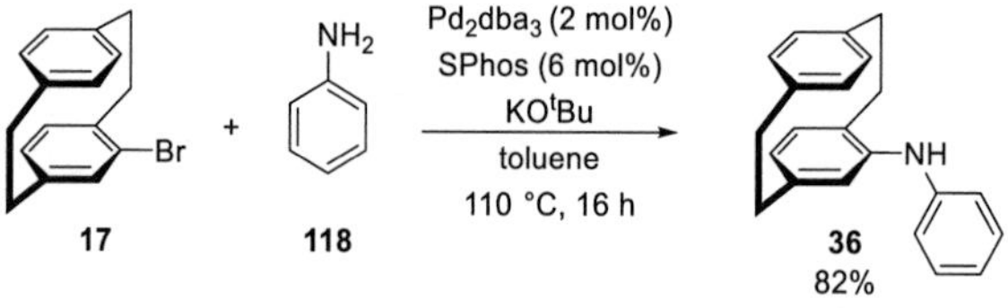

Scheme 58. Hartwig-Buchwald cross-coupling protocol towards *N*-phenylamine **36**.

The final oxidative cyclization to carbazolophane **37** was further screened starting from the conditions reported by Bolm *et al.*[39] since no dry oxygen, but only pressurized air was available. For this, a Teflon tube was inserted in the boiling mixture to constantly aerate the reaction. All reactions were carried out at a 0.1 M concentration and monitored by TLC until complete consumption of the starting material was observed. The reported conditions used compressed air, 20 mol% of Pd(OAc)$_2$ and 60 mol% of pivalic acid to obtain a yield of 62% on a 400 μmol scale (74 mg). As the synthetic core of this topic was to obtain large amounts of material, the following screening was conducted on larger scales (Table 22). Pivalic acid was added to the reaction solely in order to keep the active palladium(II) species in solution, the first two attempts excluded pivalic acid. While an equimolar amount of the catalyst yielded 46% (entry 1), 50 mol% of the catalyst gave significantly lower yields of 15%. A further decrease of the catalyst loading to 30 mol% with parallel addition of 90 mol% pivalic acid (entry 3) increased the yield to 28% when compared to entry 2. Keeping the catalyst loading at 50 mol% and a threefold excess of pivalic acid led to an increase of the yield to 49% (entries 4–5), which is significantly better than without the additive (entry 2). Decreasing the catalyst to additive ratio from 1:3 to 1:2 (entry 7) again leads to a drop in yield, another influence on performance appeared to be the scale of the reaction, when comparing entries 5 to 6, the yield almost halved when the scale is increased by a factor of ten. As the catalyst loading of 50 mol% is a significant cost factor on a multigram scale, a further decrease of the catalyst and additive loading to 25 mol%/75 mol% was tested (entries 8–11). The yields remained low in a range of 18–26% which is relatively constant for a multigram (3.40 g–7.96 g) scale.

Table 11. Screening conditions of oxidative cyclization towards carbazolophane **37**.

Entry	Scale [mmol]	Catalyst [mol%]	Additive [mol%]	Isolated yield [%]
1	0.50	100 mol%	–	46% (69 mg)
2	3.50	50 mol%	–	15% (115 mg)
3	8.00	30 mol%	90 mol%	28% (665 mg)
4	1.10	50 mol%	150 mol%	49% (161 mg)
5	1.50	50 mol%	150 mol%	49% (220 mg)
6	13.7	50 mol%	150 mol%	27% (1.09 g)
7	15.0	50 mol%	100 mol%	18% (814 mg)
8	11.5	25 mol%	75 mol%	18% (604 mg)
9	13.5	25 mol%	75 mol%	23% (910 mg)
10	21.5	25 mol%	75 mol%	23% (1.48 g)
11	27.0	25 mol%	75 mol%	26% (2.07 g)

In summary, a significant drop in isolated yield was observed when attempting to upscale the reaction. It is noteworthy, that all reactions were quenched only after full consumption of the starting material which occurred within 10 h. No spot other than the desired product was observed on TLC, but because of the huge catalyst loading and the acidic reaction conditions in acetic acid, workup was tedious in terms of removing all metal precipitate. Polymeric side products supposedly formed which could not be isolated.

From previous work, it was known that the additional steric bulk of the [2.2]paracyclophane significantly impedes, or even prohibits reactions such as Hartwig-Buchwald cross-coupling on the secondary amine functionality of **36**[126] or with inverted reactivity for e.g. a secondary amine to bromo[2.2]paracyclophane.[147] To answer this question a swift comparison of NH reactivity in nucleophilic aromatic substitution of electron poor fluoride **123** was conducted (Scheme 59). Interestingly, the secondary amine **36** did not show any conversion while the carbazolophane **37** gave the desired product in moderate yields. As the reaction proceeds *via* a nucleophilic intermediate which is generated by the basic conditions, the pK_a values of the respective amines can be discussed. Since no measured values are available, an estimate based on the literature known values in DMSO of the diphenylamine (pK_a 25)[148] and the carbazole (pK_a 20) can be taken.[149] Surely the actual values for **36** and **37** are increased due to the additional electron density of the [2.2]paracyclophane moiety. Since the pK_a value of HPO$_4^{2-}$ is only reported in water (pK_a 12.3) and an increase of this value is generally observed in DMSO,[150] it is possible that the diarylamine **36** cannot be efficiently deprotonated. Another factor to be considered is the difference in reactivity caused by the geometric differences of the amines. The crystal structure of diphenylamine shows an angle of 49° between the two phenyl planes,[151] a similar behavior is expected for **36**. As the phenyl group is rotationally free, it will twist away from the adjacent ethyl bridge and the second deck of the [2.2]cyclophane moiety, leaving the NH function rather sterically shielded between the second deck of the [2.2]paracyclophane, the ethyl bridge and the phenyl group. In contrast to that, the carbazolophane **37** is fixed in the orientation of the NH function as being part of the carbazole plane. Although the ethyl bridge is in *ortho*-position, a relatively unhindered attack can occur from the bottom of the carbazole plane.

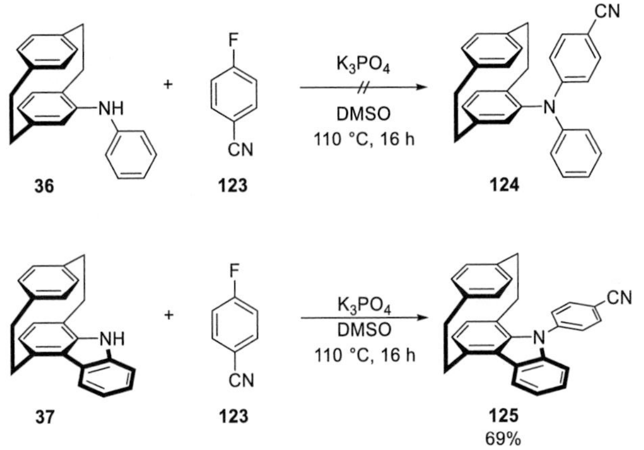

Scheme 59. Comparison of S$_N$Ar reactivity of a cyclophanic diphenylamino group in **36** and an aromatic amino group in **37**.

Motivated by the promising results for S$_N$Ar performance of **37**, the synthesis of emitter **117** was conducted under identical conditions to furnish the desired product in a moderate yield of 52% (Scheme 60).

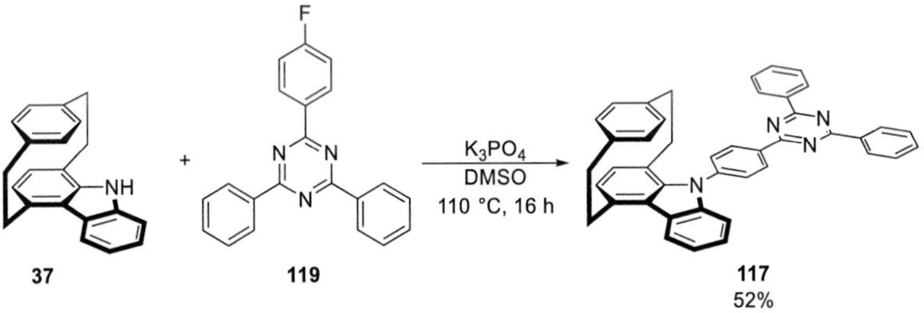

Scheme 60. SNAr towards the first carbazolophane-based emitter **117**.

A crystal structure could be obtained for emitter **117** (Figure 33), showing that the angle between the central pyrrol plane of the carbazolophane donor and the bridging phenyl plane is 55.9°, which is an increase of 10.8° when compared to the emitter **76**. In contrast, the angle between the triazine plane and the bridging phenyl is 4.9°, which is very similar to the 5.51° observed for emitter **76**. From a structural point of view, the introduction of the carbazolophane donor unit **37** did introduce a significant additional twist of 55.9°, which is in a promising range, since both a fully parallel and a fully orthogonal relative orientation quench the TADF performance either *via* overlapping of the HOMO and LUMO, or a decrease of the oscillation strength.

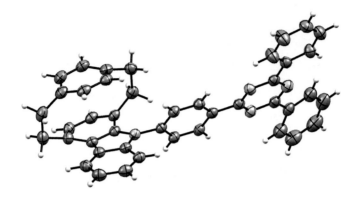

Figure 33. Molecular structure of carbazolophane emitter **117** drawn at 50% probability level.

3.2.1.4   Synthesis of Extended Carbazolophanyl Emitters

In order to further broaden the scope of carbazophanyl donor-based emitters, two derivatives were prepared starting from 4-functionalized anilines. A 4-*N*-carbazolyl amine **128** was implemented to increase the delocalization of the HOMO unit, while a 4-*tert*-butyl aniline (**132**) was selected to increase solubility for potential multidonor TADF systems.

First the 4-*N*-carbazolyl amine **128** was synthesized in a two-step procedure (Scheme 61) *via* nucleophilic aromatic substitution of 4-fluoro-nitrobenzene (**126**) with carbazole (**122**) and subsequent reduction of the nitro group of intermediate **127** using tin(II)chloride in 57% over two steps.

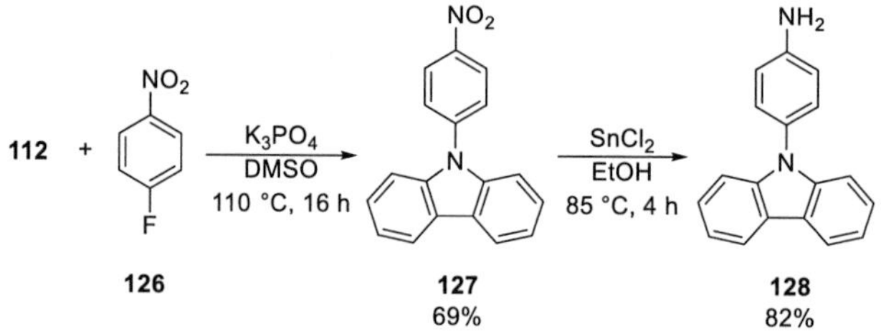

Scheme 61. Synthesis of the 4-*N*-carbazolyl aniline **128**.

The secondary amino-[2.2]paracyclophane intermediate **129** was obtained using identical conditions as for **36** (see Scheme 58). Unfortunately, the reaction performed poorly on a 7 g scale yielding only 12% (Scheme 62). Despite this setback, enough material was isolated to further proceed with the synthesis.

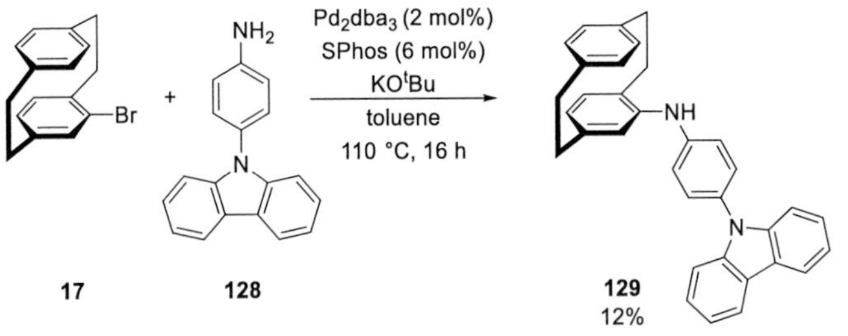

Scheme 62. Synthesis of *N*-carbazolyl extended *N*-phenylamino[2.2]paracyclophane **129**.

With the results of the oxidative cyclization screening of Table 11 in mind, a small-scale synthesis was chosen with a slightly reduced catalyst loading of 20 mol% using 560 mg of starting material (Scheme 63). The extended carbazolophane donor **130** was obtained in 34% yield.

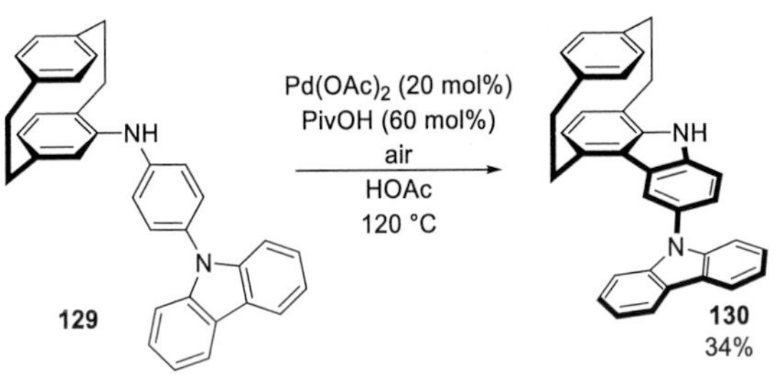

Scheme 63. Oxidative cyclization to the *N*-carbazolyl extended donor **130**.

Finally, the extended donor unit **130** was connected to the triphenyltriazine acceptor *via* nucleophilic aromatic substitution in an identical protocol as for the derivatives described before. The extended TADF emitter **131** was isolated in 37% yield (Scheme 64).

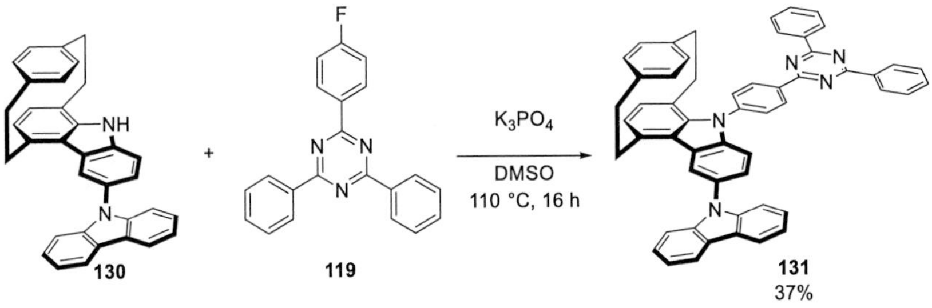

Scheme 64. Final step of the synthesis of the extended emitter **131**.

Unfortunately, no suitable single crystals could be obtained for crystal structure analysis.

As mentioned before, a second derivative was targeted featuring a *tert*-butyl group for increased solubility. Hartwig-Buchwald cross-coupling was performed with commercially available 4-*tert*-butyl-aniline (**132**) giving the desired intermediate in 47% yield (Scheme 65).

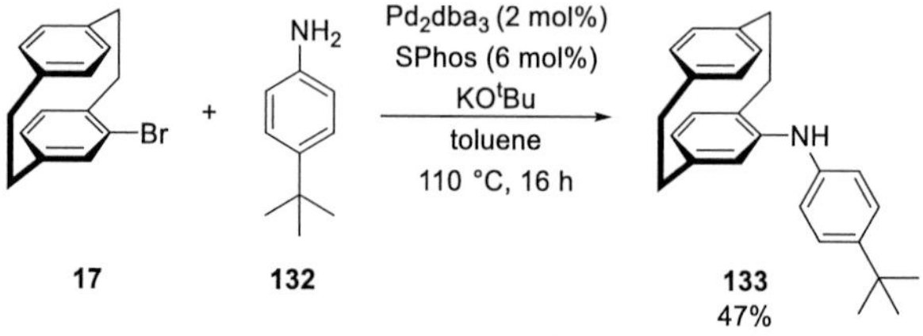

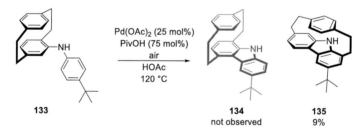

Scheme 65. Synthesis of the *tert*-butyl-extended diphenylamine **133**.

Intriguingly, the following oxidative cyclization of **133** did not yield the desired product, but the regio isomer **135** in poor yield (Scheme 66). Following this oxidative cyclization, a full consumption of the starting material was observed on TLC with only one new and clear spot forming in the expected eluent polarity range of *n*-hexane/ethyl acetate (5:1). Therefore, it can be safely assumed that the expected carbazolophane **134** did not form. The significant loss of material is commonly observed for the oxidative cyclization protocols due to the formation of oligo- and polymeric side products or palladium complexes.

Scheme 66. Unexpected reactivity upon attempted oxidative cyclization of **133** yielding a (1,7)-bridged carbazolophane **135**.

The structure of **135** was unambiguously verified by single crystal analysis (Figure 34). This intriguing hitherto unreported structure is a [2]paracyclo[2](1,7)carbazolophane featuring a *tert*-butyl group in position 3 of the carbazole subunit. Numerous structural features are found in this new cyclophane. I) The two phane planes are not perfectly parallel as indicated by the distances of each pyrrole atom to the xylene plane. The distance varies between 2.97–3.45 Å with the shortest distance obtained at the quaternary carbon between C1 and the nitrogen atom. This distance range is identical to the [2.2]paracyclophane (see chapter 1.1.1). II)The *para*-xylene unit does not bend, but all six aromatic carbon atoms lie in one plane. The methenyl groups also lie within that plane, or are unsignificantly bent out of the xylene plane. This can be reasoned by the increased size of the (1,7)-bridged carbazolophane. III) The increased distance between the bridgehead positions at the carbazole demands the carbazole moiety to be bent within the aromatic structure. An angle of 38° is measured between the two benzene planes of the carbazole unit (Figure 34C), which raises the question about the aromaticity of the carbazole unit in this (1,7)carbazolophane.

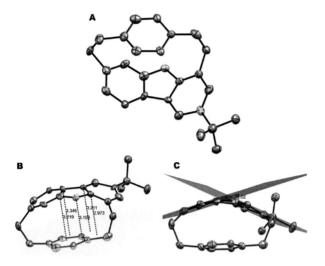

Figure 34. Molecular structure of the (1,7)carbazolophane **135** obtained by single crystal analysis, drawn at 50% probability level. A) Overview; B) Distances of the pyrrole moiety to the xylene plane; C) Angle between the two benzene moieties of the highly strained carbazole unit.

One insight on aromaticity was given by ^{1}H NMR spectroscopy. Since the carbazolophane **135** is bridged in unsymmetrical positions, the second xylene deck is only partially on top of the carbazole (Figure 34B). This leads to a significant upfield shift of only two of the four aromatic xylene hydrogens. When looking at the aromatic region of the ^{1}H NMR (Figure 35), two signals are observed at 5.20 ppm and 4.46 ppm. As the xylene moiety is not bent, this unexpectedly strong shielding can only be explained by the aromatic ring current of the pyrrole/carbazole moiety. For comparison, the upfield shift of the analogous xylene atoms in **37** only reach 5.94 ppm and 5.22 ppm.

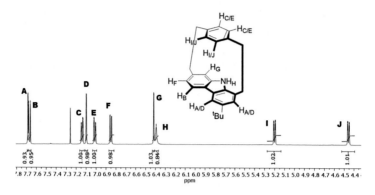

Figure 35. Aromatic ^{1}H NMR region of **135**, showing the significant upfield shift of the stacked xylene unit for signals I and J.

As the novel (1,7)-bridged carbazolophane **135** was only obtained in small amounts, no conversion to a TADF emitter could be performed. Despite this, the mechanism leading to this new class of cyclophanes should be discussed (Scheme 67). As described in the introduction (see chapter 1.1.7), the bridge-head carbon atom is susceptible to an electrophilic attack by an electron deficient alkene[54] or a cumulene.[56] In contrast to those two known examples, the mechanism present here is palladium(II) mediated. Yet, palladium(II) species do react as electrophiles as well. Clearly the amine increases the electron density of all *ortho* (and *para*) positions. In case of **133**, these positions are the *ortho*-C–H of the [2.2]paracyclophane *o*-C–H^P, the *ortho*-quaternary bridge-head carbon *o*-C^q, and the two equivalent *ortho*-C–H of the aniline group indicated with *o*-C–H^A. The *para*-positions are in principle possible, but an addition to these positions will not lead to the desired product. For this reason, *o*-H^P can be excluded, as well. An palladium(II)-mediated oxidative addition to a quaternary carbon center as for *o*-C^q is uncommon, but has to occur during the reaction. Therefore, the initial step of oxidative addition can proceed at the aniline moiety leading to **136**, or **137**. As oxidative addition is much more common for a primary sp^2-center and there are two equivalent *o*-C–H^A present in the starting material, intermediate **136** is much more likely. Nevertheless, both possible intermediates undergo a second oxidative addition at the respective complementary position yielding **138** as the key intermediate of this mechanism. It is noteworthy, that the **136** intermediate would be still able to cyclize at the *o*-C–H^P position yielding the initially expected (1,4)-bridged carbazolophane, while the **137** intermediate is only able to cyclize to the key intermediate **138**.

After the reductive elimination to **139**, a sigmatropic [1,5]-rearrangement takes place, shifting the ethyl bridge to the remaining *ortho*-C–H of the aniline to yield the final product **135**.

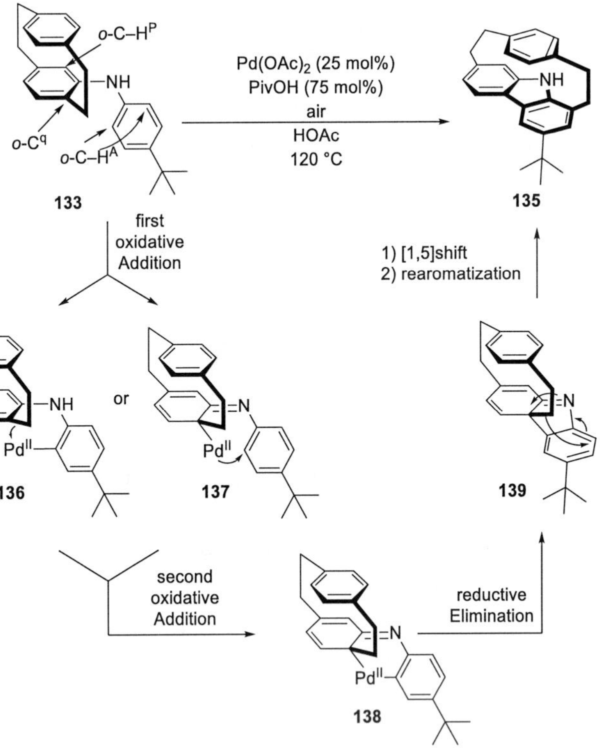

Scheme 67. Proposed mechanism towards (1,7)-bridged carbazolophane **135**.

Electronic effects are to be taken into account when discussing why the *tert*-butyl extended amine **133** and the non-extended amine **36** react with different regioselectivities under identical conditions. For this a closed look has to be given to the initial oxidative cyclization (Scheme 68). This initial oxidative addition would generally proceed at the most electron-rich position. In most cases, this would be the [2.2]paracyclophane position o-C–H^P yielding intermediate **141**. While this is true for the non-substituted aniline derivative **36**, the selectivity may change for the *para-tert*-butyl derivative **133**. Intriguingly, upon addition to the o-C–H^A positions leading to **142** and/or **143** both products (**147** and **148**) are still mechanistically available. Although intermediates **142** and **143** are structurally identical, they mayt not be from a sterical point of view caused by the hindered rotation of the complex by the ethyl bridges and the inability of complex **143** to cyclize at the quaternary position due to the same steric hindrance. Therefore, there must be an additional stereoelectronic influence to propel a stereoselective addition at the quaternary carbon leading to complex **145**.

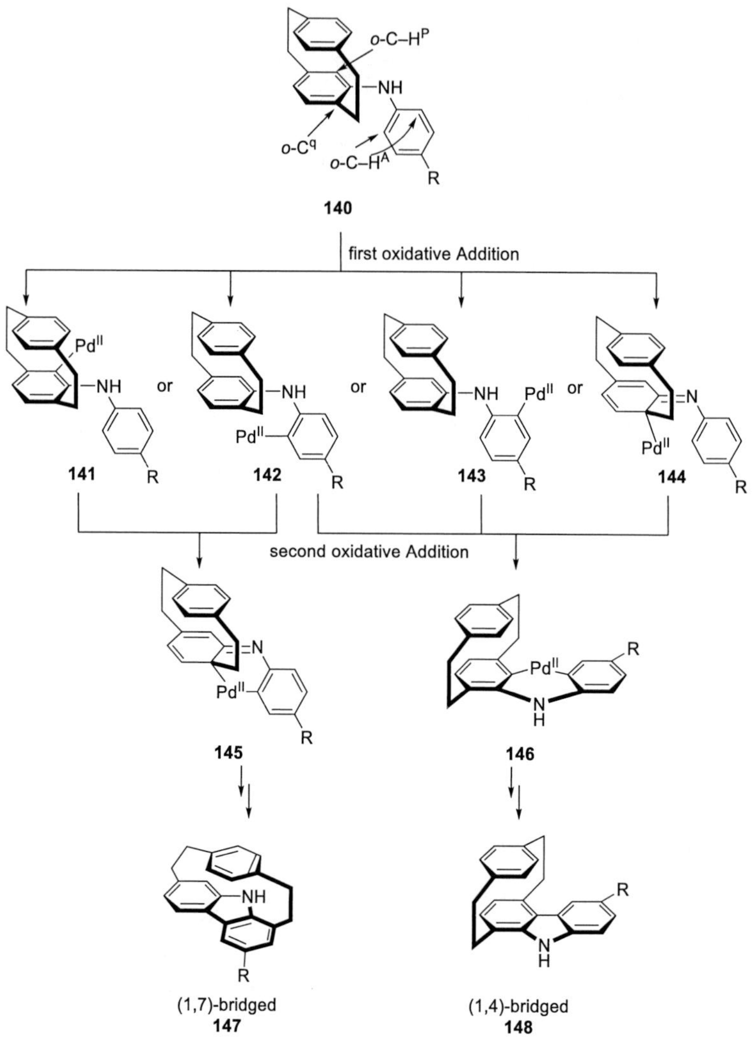

Scheme 68. Overview over possible reaction pathways after the first oxidative addition towards carbazolophanes. Stereo configuration descriptors were omitted for clarity.

Compared to both literature and own results obtained during supervised preliminary work (Figure 36),[144] a switch of stereoselectivity caused by an electron-rich aniline derivative cannot be the only explanation under identical reaction conditions, as an isopropyl and methyl derivative reported in literature[39] and the electron-rich N-phenoxazine derivative **149** prepared under identical conditions as discussed in this thesis do yield the expected (1,4)-bridged carbazolophanes.[144]

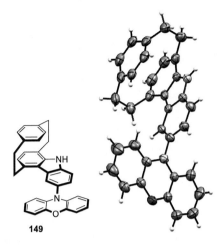

149

Figure 36. Molecular structure of a phenoxazine-extended (1,4)-bridged carbazolophane donor **149** drawn at 50% proability level.[144]

In summary, an oxidative addition to the aniline does happen in any case, while the competitive addition to the o-C–H^p vs. o-C^q positions of the [2.2]paracyclophane moiety is the cause of selectivity. A thorough insight on this [2.2]paracyclophane specific behavior can only be gained by further experimental and theoretical work.

Finally, to test the reactivity of the carbazolophane **37**, and to simplify a potential access to a wider scope of derivatives, a bromo substituent was introduced with NBS (Scheme 69) in 77% yield. Subsequent homo-coupling towards a donor-dimer showed not to be successful. In order to keep the thesis concise, this route was not further explored.

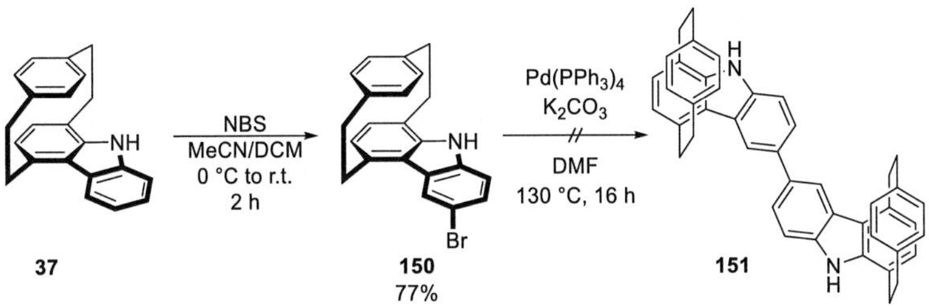

Scheme 69. Bromination and subsequent unsuccessful homo-coupling of the free carbazolophane **37**.

3.2.2   Characterization

With the parent emitter **76**, the carbazolophane emitter **117** and the *N*-carbazolyl-extended carbazolophane emitter **131** in hand (Figure 37), sublimation, calculations, electrochemical, spectroscopic and device analyses were performed in collaboration with the Zysman-Colman group in St. Andrews, Scotland. All experiments in this chapter were conducted by Dr. Nidhi Sharma.

Both the parent emitter **76** and the carbazolophane emitter **117** could be successfully purified by train sublimation. Unfortunately, the *N*-carbazolyl-extended emitter **131** decomposed, therefore the main discussion will focus on **76** and **117**. Additionally, a more material based nomenclature introduced in Figure 37 will be used in this chapter.

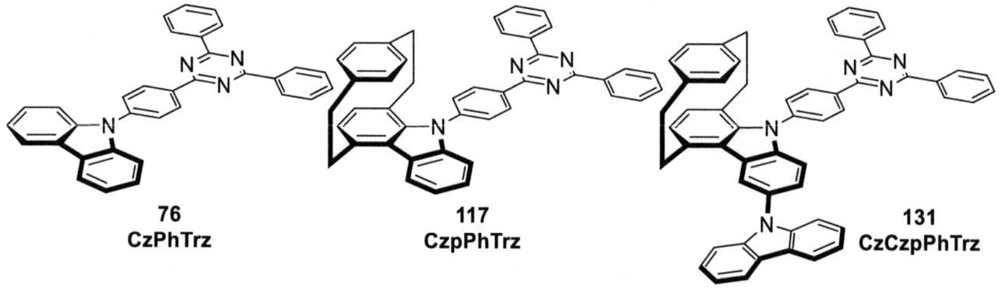

Figure 37. Overview of the evaluated emitters with the material-based nomenclature.

3.2.2.1   DFT Calculations

DFT calculations were performed with the Gaussian 09 revision D.018 suite.[130] Initially the geometries in the ground state in gas phase were optimized employing the PBE0[131] functional with the standard Pople 6-31G(d,p) basis set.[132] Time-dependent DFT calculations were performed using the Tamm–Dancoff approximation (TDA).[133] The molecular orbitals were visualized using GaussView 5.0 (Figure 38) software to get initial insight on the nature of the tuned properties induced by the carbazolophane donor structure. In all emitters, the HOMO is localized mainly on the donor unit with decreasing contribution on the linking phenyl unit with increasing donor size, while the LUMO is homogeneously distributed over all three phenyl rings of the triphenyltriazine acceptor unit. While for **CzpPhTrz**, the HOMO is significantly extended onto the second [2.2]paracyclophane deck, presumably because of the strong π–π through-space interactions, a different distribution is seen for **CzCzpPhTrz**. In the *N*-extended donor, no contribution of the second deck is seen, while the majority of the HOMO is located at the additional carbazole, which is almost perpendicular to the carbazolophane subunit. The twist at the C–N bond in **CzPhTrz** is calculated to be 51°, which is in good agreement with the 45° determined by single crystal analysis and the large ΔE_{ST} of 0.38 eV, which is in accordance with previous reports of Lee *et al.* (ΔE_{ST} of 0.36 eV).[152] For **CzpPhTrz**, the torsion is increased to 58°, which is in accordance with the result of the single crystal analysis (56°). Caused by the extended conjugation, the carbazolophane unit in **CzpPhTrz** is a stronger donor, which results in a smaller ΔE_{ST} of 0.30 eV and a lower S_1 energy of 3.11 eV. That same trend is even more pronounced for **CzCzpPhTrz** with a ΔE_{ST} of 0.20 eV and an S_1 energy of 2.98 eV.

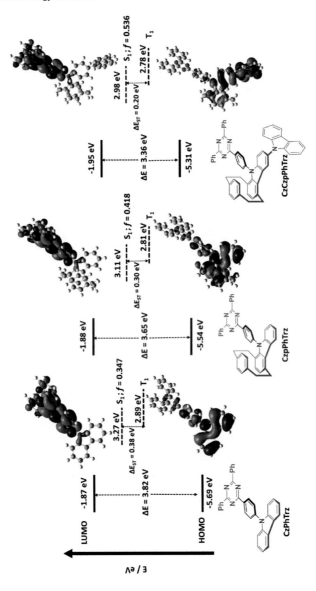

Figure 38. DFT calculations of the three prepared emitters. Results and images provided by Dr. Sharma.[134]

3.2.2.2   Physical Properties

After the successful sublimation of emitters **76** (**CzPhTrz**) and **117** (**CzpPhTrz**), a thermal characterication was conducted with these two emitters *via* thermal gravimetric analysis (TGA) and differential scanning calorimetry (DSC) as shown in Figure 39. Both emitters are stable up to 300 °C – which is needed for a sublimation-based OLED preparation to be successful – before a significant mass loss occurs, and the melting points determined from DSC show that both emitters do not undergo any calorimetrically visible process, such as melting or in the case of the [2.2]paracyclophane scaffold an isomerization which occurs above 260 °C. As emitter **131** (**CzCzpPhTrz**) proved to be unsuitable for sublimation, it was not included in further investigations since it would not have been possible to obtain sublimed OLED devices from a pure emitter.

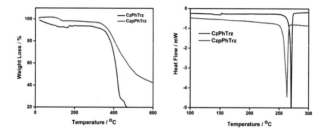

Figure 39. TGA (left) and DSC (right) measurements of **CzPhTrz** and **CzpPhTrz**. Results and images provided by Dr. Sharma.[134]

Determining the HOMO and LUMO levels of **CzPhTrz** (ΔE_{redox} = 2.86 eV) and **CzpPhTrz** (ΔE_{redox} = 2.58 eV) by cyclic voltammetry showed that in accordance with the theoretical predictions, the electrochemical HOMO-LUMO gap of emitter **CzpPhTrz** is smaller due to a reduced ionization potential of the donor (for summary see Table 13). The DFT predictions are further verified by UV-Vis absorption and photoluminescence (PL) spectra as seen in Figure 40 with the bathochromic shift for **CzpPhTrz** (λ_{PL} = 470 nm) compared to **CzPhTrz** (λ_{PL} = 446 nm). Both compounds exhibited unstructured PL spectra in toluene, indicative of an excited state with strong intramolecular charge transfer (ICT) character. The Φ_{PL} values for **CzPhTrz** and **CzpPhTrz** are 60% and 70%, which decreased to 58% and 55%, respectively, upon exposure to air. This increase of Φ_{PL} for **CzpPhTrz** is consistent with the higher predicted oscillator strength *f*.

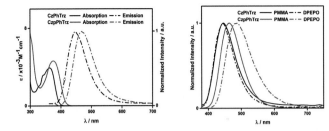

Figure 40. Left: UV-Vis absorption and PL spectra in toluene, Right: PL spectra of DPEPO thin films at λ_{exc} = 360 nm. Results and images provided by Dr. Sharma.[134]

Next, the PL behavior in the solid-state as doped thin films was evaluated in Table 12. For this the non-sublimed emitter **CzCzpPhTrz** could be included for spin-coated films. When comparing the neat films to one another (entry 1), the predictions from DFT match the experimental data showing an increasing quantum yield (Φ_{PL}) with increased donor size which is a result of the oscillator strength. Comparing spin-coated PMMA films (entry 2), the quantum yields for **CzPhTrz** and **CzCzpPhTrz** do not increase significantly, while for **CzpPhTrz**, Φ_{PL} reaches 45%. The bathochromic shift of the emission wavelength is linked to the increased donor from **CzPhTrz** (446 nm), over **CzpPhTrz** (464 nm) to **CzCzpPhTrz** (480 nm). A further increase of quantum yield is observed in DPEPO (entries 3–4), accompanied by additional red-shift caused by the polar host. Optimization of **CzpPhTrz** in mCP (entry 5) and CzSi (entry 6) do not show an improvement over DPEPO (entry 4). A spin-coated film of **CzCzpPhTrz** in in mCP (entry 5) showed an excellent quantum yield of 65% with an emission peak at 486 nm.

Table 12. Screening of different host materials for optimal photoluminescence quantum yields. Results and images provided by Dr. Sharma.[134]

Entry	Host Material[a]	CzPhTrz	CzpPhTrz	CzCzpPhTrz
		Φ_{PL} [%]	Φ_{PL} [%]	Φ_{PL} [%]
		(λ_{PL} [nm])	(λ_{PL} [nm]	(λ_{PL} [nm]
1	Neat Film	20	28 (470)	40[b] (476)
2	10 wt. % PMMA[b]	25 (446)	45 (464)	48 (480)
3	1 wt.% DPEPO	39 (448)	43 (482)	–
4	10 wt.% DPEPO	50 (448)	69 (482)	–
5	15 wt.% mCP	–	62 (465)	65[b] (486)
6	10 wt.% CzSi	–	60 (464)	–

[a] Thin films were prepared by vacuum deposition and values were determined using an integrating sphere (λ_{exc} = 360 nm); degassing was done by N_2 purge. – [b] Prepared by spin coating.

Photoluminescence lifetime measurements of **CzPhTrz** and **CzpPhTrz** were conducted in degassed toluene (Figure 41). In this unipolar solvent, the emission decay times are monoexponential and in the nanosecond regime, indicating, that no delayed fluorescence behavior is present. The absence of delayed fluorescence was also assumed for the unipolar PMMA host.

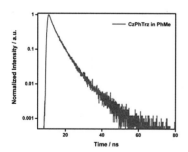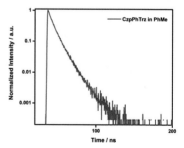

Figure 41. Photoluminescence decay curves of **CzPhTrz** and **CzpPhTrz** in degassed PhMe (λ_{exc} = 378 nm). Results and images provided by Dr. Sharma.[134]

Therefore, 10 wt% DPEPO thin-films were tested to show that still no delayed component is observed for the parent emitter **CzPhTrz** (Figure 42), which is in accordance with literature,[152] while a non-monoexponential microsecond emission lifetime is seen for the carbazolophane derivative **CzpPhTrz** (see Figure 43).

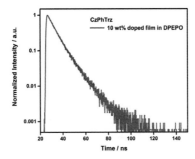

Figure 42. Photoluminescence decay curve of **CzPhTrz** in 10 wt.% DPEPO thin films (λ_{exc} = 378 nm). Results and images provided by Dr. Sharma.[134]

For **CzpPhTrz**, both prompt (τ_p = 7 ns) and delayed (τ_d = 65 μs) emission components were observed (Figure 43, left). The temperature dependence of the delayed lifetime of **CzpPhTrz** (Figure 43, right) with increasing delayed emission components additionally verify the TADF characteristics of the newly designed emitter.

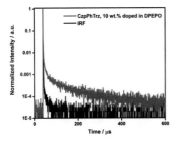 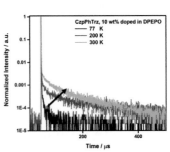

Figure 43. Left: Photoluminescence decay curve of **CzpPhTrz** in 10 wt% DPEPO thin films. Right: Temperature dependence of photoluminescence decay curve of **CzpPhTrz** in 10 wt.% DPEPO thin-film (λ_{exc} = 378 nm), IRF = Instrument Response Function. Results and images provided by Dr. Sharma.[134]

Furthermore, ΔE_{ST} of **CzPhTrz** (0.32 eV) and **CzpPhTrz** (0.16 eV) were determined from the singlet and triplet energies estimated from the onset of the prompt and delayed emission spectra, measured in 10 wt.% DPEPO doped films at 77 K (Figure 44).

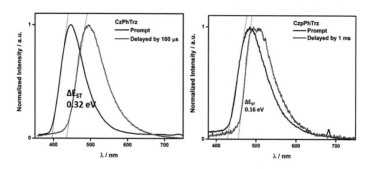

Figure 44. Prompt and delayed spectra at 77 K of **CzPhTrz** (λ_{exc} = 378 nm) **CzpPhTrz** (λ_{exc} = 378 nm).in 10 wt.% DPEPO film. Results and images provided by Dr. Sharma.[134]

There results clearly show that the introduction of the bulky [2.2]paracyclophane moiety within the donor structure not only turns on the TADF emission channel, but it is also possible to maintain a high luminescence quantum yield.

All electrochemical and photophysical characterization is summarized in Table 13.

Table 13. Optical Properties of **CzPhTrz** and **CzpPhTrz**. Results and images provided by Dr. Sharma.[134]

Entry	Emitter	λ_{abs} [a) [nm]	λ_{PL} [a) [nm]	λ_{PL} [c) [nm]	Φ_{PL} [b) [%]	Φ_{PL} [c) [%]	τ_p [c) [ns]	τ_d [c) [µs]
1	**CzPhTrz (76)**	363	446	448	60 (58)	50	4.8	–
2	**CzpPhTrz (117)**	375	470	482	70 (55)	69	7	65

		HOMO [d),e) [eV]	LUMO [d),e) [eV]	ΔE_{redox} [f) [eV]	S_1 / T_1 [g) [eV]	ΔE_{ST} [g) [eV]
1	**CzPrTrz (76)**	−6.11	−3.26	2.86	3.14 / 2.82	0.32
2	**CzpPhTrz (117)**	−5.80	−3.23	2.58	2.89 / 2.80	0.16

[a) In PhMe at 298 K. – [b) Quinine sulfate (0.5 M) in H_2SO_4 (aq) was used as reference (Φ_{PL}: 54.6%, λ_{exc} = 360 nm).[137] Values quoted are in degassed solutions, which were prepared by three freeze-pump-thaw cycles. Values in parentheses are for aerated solutions, which were prepared by bubbling air for 10 min. – [c) Thin films were prepared by vacuum depositing 10 wt.% doped samples in DPEPO and values were determined using an integrating sphere (λ_{exc} = 360 nm); degassing was done by N_2 purge for 10 minutes. – [d) In DCM with 0.1 M [nBu_4N]PF_6 as the supporting electrolyte and Fc/Fc^+ as the internal reference (0.46 V vs. SCE).[153] – [e)The HOMO and LUMO energies were determined using $E_{HOMO/LUMO} = -(E^{ox}_{pa,1}/ E^{red}_{pc,1}+ 4.8)$ eV[154] where E^{ox}_{pa} and E^{red}_{pc} are anodic and cathodic peak potentials, respectively. – [f) $\Delta E_{redox} = |E_{HOMO}-E_{LUMO}|$. [g) Determined from the onset of prompt and delayed spectra of 10 wt.% doped films in DPEPO, measured at 77 K (λ_{exc} = 355 nm).

3.2.3 Device Performance

Based on the promising characterization and the verification of TADF for **CzpPhTrz** in DPEPO ((bis[2-(diphenylphosphino)phenyl]ether oxide)) films, vacuum-deposited OLEDs were fabricated for **CzPhTrz** and **CzpPhTrz**. The device architecture was chosen in a way that DPEPO remained the host of the emissive layer (EML). As DPEPO possesses a HOMO energy of -6.3 eV[155] and the excitons have to be confined within the EML, mCP ((1,3-bis(N-carbazolyl)benzene)) and DPEPO were chosen as electron (EBL) and hole blocking layers (HBL). The electron transporting layer (ETL) was comprised of 1,3,5-tris(3-pyridyl-3-phenyl)benzene (TmPyPB), which possesses a high electron mobility of 10^{-4} cm^2 V^{-1} s^{-1} and a high triplet energy of 2.75 eV along with a deep HOMO energy of 6.7 eV.[156] N,N'-bis(naphthalen-1-yl)-N,N'-bis(phenyl)benzidine (NPB) was used as hole injection layer (HIL) and tris(4-carbazoyl-9-ylphenyl)amine (TCTA) was used as hole transporting layer (HTL) on an ITO substrate and anode with a LiF/Al cathode. The resulting device architecture for both CzPhTrz and CzpPhTrz was ITO/ NPB (30 nm)/ TCTA (20 nm)/ mCP (10 nm)/ **Emitter**:DPEPO (10 wt.%, 20 nm)/ DPEPO (10 nm)/ TmPyPB (40 nm)/ LiF (1 nm)/ Al (100 nm) as shown in Figure 45.

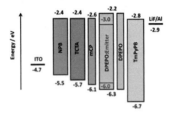

Figure 45. Chemical structures and energy levels of materials used for the device fabrication. Results and images provided by Dr. Sharma.[134]

The electroluminescence (EL) properties of the devices are summarized in Figure 46 and Table 14. OLEDs based on **CzPhTrz** showed a deep blue emission with a λ_{EL} of 446 nm and CIE (commission international de l'eclairage) coordinates of (0.16, 0.12) whereas devices using **CzpPhTrz** showed an expected red-shifted emission with a sky blue λ_{EL} of 482 nm and CIE coordinates of (0.17, 0.25). Both devices exhibited steep current-voltage-luminance behavior and low turn-on voltages (V_{on}) of 3.6 V (**CzPhTrz**) and 3.2 V (**CzpPhTrz**). The lower V_{on} for the OLED with **CzpPhTrz** is attributed to the lower energy gap of this emitter.

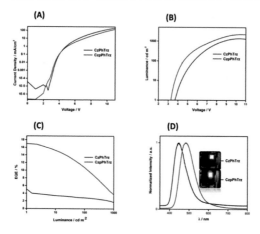

Figure 46. (A) Current density-voltage characteristics, (B) Luminance vs. Voltage, (C) EQE vs Luminance, (D) Normalized EL spectra of **CzPhTrz** and **CzpPhTrz**. Results and images provided by Dr. Sharma.[134]

The external quantum efficiency (EQE) vs. luminance of the two OLEDs shown in Figure 46C exhibits a high EQE of 17% for the TADF emitter **CzpPhTrz** at low brightness (1 cd m^{-2}). The efficiency roll-off is rather low with an EQE$_{100}$ (at 100 cd m^{-2}) of 12%. This is accompanied by a maximum current efficiency (CE$_{max}$) of 34.8 cd A^{-1} and a maximum power efficiency (PE$_{max}$) of 32.5 lm W^{-1}. In contrast, devices with **CzPhTrz** showed an EQE$_{max}$ of 5.8% and an EQE$_{100}$ of 2.8%. This poor performance is caused by the inability of the emitter to harvest electrically generated triplets *via* TADF mechanism. In summary, a threefold enhancement of EQE$_{max}$ and a fourfold enhancement of the display relevant brightness EQE$_{100}$ was observed by introduction of the carbazolophane unit **37** into the design of a TADF emitter. Additionally, **CzpPhTrz** is amongst the state-of-the art blue to sky-blue TADF triazine-based emitters in terms of EQE$_{max}$.

Table 14. Electroluminescence properties of **CzPhTrz** and **CzpPhTrz**. Results and images provided by Dr. Sharma.[134]

Entry	Emitter	V$_{on}$ [a] [V]	λ$_{EL}$ [b] [nm]	EQE$_{max}$ [c] ; EQE$_{100}$ [d] [%]	CE$_{max}$ [c] [cd A^{-1}]	PE$_{max}$ [c] [lm W^{-1}]	CIE [e] (x,y)
1	**CzPhTrz (76)**	3.6	446	5.8 ; 2.8	10.5	8.6	(0.14, 0.12)
2	**CzpPhTrz (117)**	3.2	480	17.0 ; 12.0	34.8	32.5	(0.17, 0.25)

[a] Measured at 1 cd/m^2. – [b] Emission maxima at 1 mA cm^{-2}. – [c] Maximum efficiencies at 1 cd m^{-2}. – [d] EQE at 100 cd m^{-2}. – [e] Commission Internationale de l'Éclairage (CIE) coordinates at 1 mA cm^{-2}.

3.2.4 Chiral Properties of Carbazolophane Emitter 117

Since the **CzpPhTrz** (**117**) emitter possesses planar chirality and showed excellent TADF performance, it was desired to obtain this emitter in an enantiopure way. As the amine group (see Scheme 59) is rather inaccessible to a chiral derivatization, a chiral HPLC separation was conducted (Scheme 70).

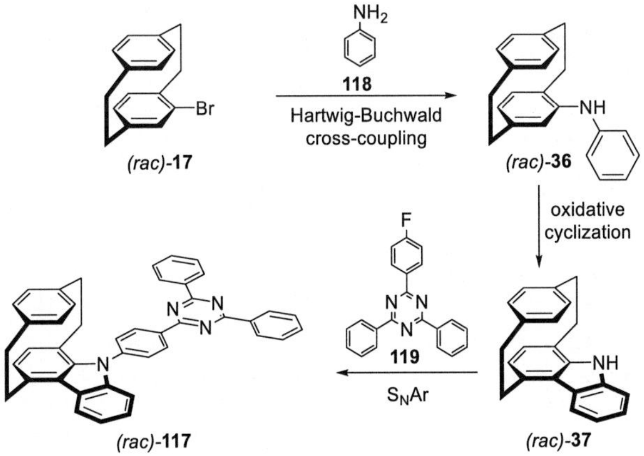

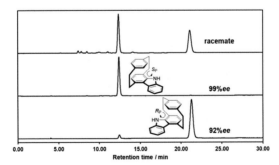

Scheme 70. Overview of the synthetic access to emitter **117**.

Since the final emitter is very unpolar, it was unsuitable for a racemate separation on a semipreparative scale, but the free carbazolophane **37** was fit. In Figure 47, the analytical HPLC chromatogram is shown to determine the enantiomeric excess, but it was also possible to separate the enantiomers on a semipreparative scale using the identical stationary phase.

Figure 47. Analytical HPLC separation of **37**. A chiral Amylose-SA column from YMC was used as stationary phase with 10% isopropanol in *n*-hexane as the eluent with a 1.0 mL/min flow rate and detection at 256 nm.

To determine the absolute configuration (result already shown in Figure 47), the literature known bromide **17** was separated into its enantiomers *via* semipreparative chiral HPLC using chiralpak-AZ as the stationary phase and acetonitrile as the eluent following group procedure.[157] The absolute configuration was determined by comparison of the circular dichroism (CD) spectrum (Figure 48) with literature-known experimental and calculated data,[158] as well as with theory.[159]

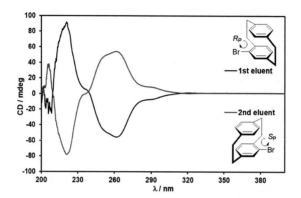

Figure 48. CD measurements in acetonitrile of 4-bromo[2.2]paracyclophane (**17**).

With these enantiomers in hand, the two-step synthesis to carbazolophane **37** was conducted and the HPLC retention times were compared of the enantioenriched carbazolophane from enantioenriched bromide **17** were compared with known chirality and the chromatographically separated carbazolophane (Figure 49 and Figure 50). For this, an analytical chiralpak-AS stationary phase was used, which gave broadened signals, but the stereoconfiguration could be unambiguously assigned. Interestingly, the first fraction eluted from amylose-SA stationary phase (Figure 47), is the second fraction eluted from chiralpak-AS stationary phase (Figure 49).

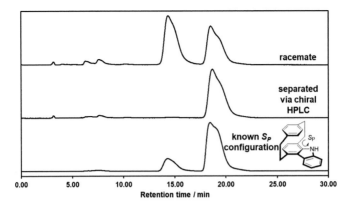

Figure 49. Analytical HPLC of (S_P)-carbazolophane **37**. Chiralpak-AS was used as the stationary phase with 5% isopropanol in n-hexane as the eluent with a 1.0 mL/min flow rate and detection at 256 nm. Top: Racemate for comparison. Middle: First fraction from Figure 47. Bottom: Matching enantiomer with known (S_P)-configuration synthesized from (S_P)-**17**.

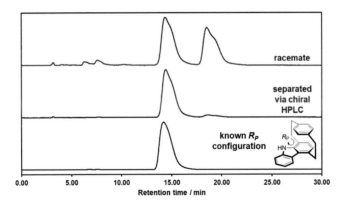

Figure 50. Analytical HPLC of (R_P)-carbazolophane **37**. Chiralpak-AS was used as the stationary phase with 5% isopropanol in n-hexane as the eluent with a 1.0 mL/min flow rate and detection at 256 nm. Top: Racemate for comparison. Middle: First fraction from Figure 47. Bottom: Matching enantiomer with known (R_P)-configuration synthesized from (R_P)-**17**.

The final conversion to the respective enantioenriched emitter **117** was then conducted with subsequent fractional recrystallization in order to increase the enantiomeric excess (Figure 51). Comparing the enantiomeric excess of (R_P)-**117** emitter (93%ee) with the carbazolophane precursor (R_P)-**37** (92%ee), the recrystallization did not serve to increase the ee in a significant manner. As the final emitter does not present pH-active groups, analytical HPLC runs had to be performed with a significantly reduced flow rate of 0.2 mL/min in order to observe a separation of the enantiomers.

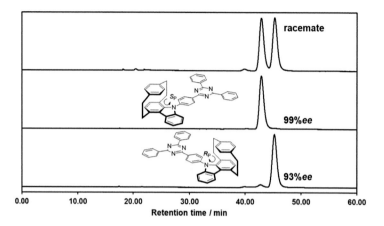

Figure 51. Analytical HPLC of (R_P)-carbazolophanyl emitters **117**. Amylose-SA was used as the stationary phase with 2% isopropanol in *n*-hexane as the eluent with a 0.2 mL/min flow rate and detection at 256 nm. Top: Racemate for comparison. Middle: (S_P)-**17** with a determined enantiomeric excess of 99%. Bottom: (R_P)-**17** with a determined enantiomeric excess of 93%.

The chiral emitters were given to Dr. Nidhi Sharma to conduct further photophysical characterizations of these compounds. Initial CD spectra (Figure 52) showed both enantiomers to be perfect mirror images of each other. The degree of CD (g_{abs} at 375 nm) was determined to be 7×10^{-3} for the R enantiomers and 5×10^{-3} for the S enantiomer.

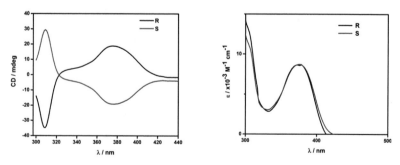

Figure 52.Circular dichroism (CD) and extinction coefficient measurements of the enantiomers. Results and images provided by Dr. Sharma.[134]

Subsequent circular polarized luminescence measurements (Figure 53) showed only a small dissymmetry factor (g_{lum} 460 nm) of 1.3×10^{-3}.

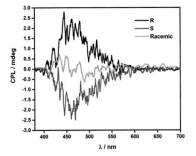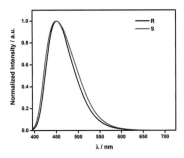

Figure 53. Circular polarized luminescence (CPL) and photoluminescence measurements of the enantiomers. Results and images provided by Dr. Sharma.[134]

When comparing to the literature, purely organic emitters TADF emitters show g_{lum} values of 1.3×10^{-3} for a BINOL-induced chiral emitter with an EQE_{max} of 9.1% at an emission wavelength of 540 nm,[160] 1.1×10^{-3} for a cyclohexyl-induced chiral emitter with an EQE_{max} of 19.8% at an emission wavelength of 520 nm,[161] and 4.24×10^{-3} for a through-space bridged [2.2]paracyclophane emitter with an emission wavelength of 525 nm in powder.[110]

In summary, the introduction of a carbazolophane donor unit to the triazine TADF structure, turned on the TADF mechanism. The resulting emitter was suitable for a state-of-the-art sky-blue OLED. Finally, the inherent planar chirality of the [2.2]paracyclophane scaffold also induced a circularly polarized luminescence which is among the bluest purely organic CPL-TADF emitter reported to date.

3.3   Unexpected *para* C–H Activation

As mentioned before (chapter 3.1.1.1, Scheme 31), the Hartwig-Buchwald cross-coupling to the *N,N*-diphenylamino[2.2]paracyclophane did not yield the desired product, as was expected based on the results of previous work.[126] Unexpectedly, a *para* C–H activation was observed and verified by single crystal analysis. This mechanistically intriguing reaction was further investigated as is shown in this chapter.

3.3.1   Initial Results

During the screening of common Hartwig-Buchwald cross-coupling conditions on the **36** model system in order to obtain diphenylamine **84**, which was a desired through-space TADF donor group (compare chapter 3.1), a mixture of two new products was obtained. Upon crystallographic analysis, none of the isolated products possessed the expected molecular structure, but a *para*-phenyl substituent (Scheme 71). Although the reaction proceeded for two days, some of the starting material (**36**) remained unconsumed.

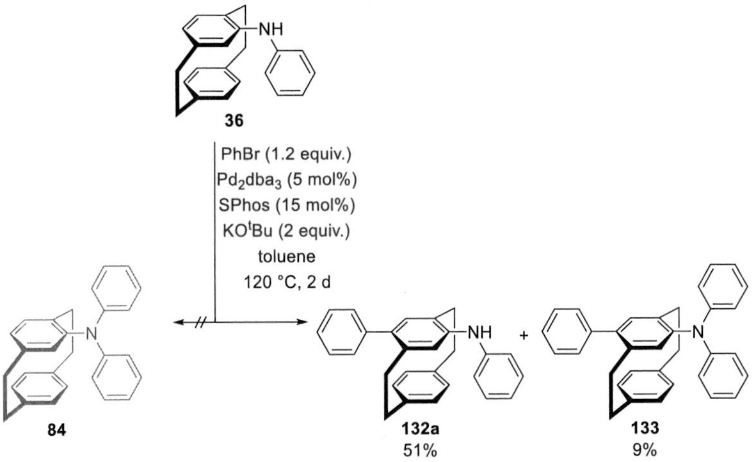

Scheme 71. Initial observation of C–H activation under standard Hartwig-Buchwald cross-coupling conditions.

Product **132a**, which was isolated in 51% yield clearly did not react at the amine position but at the *para*-position to the diphenylamine moiety in **36**. The molecular structure is shown in Figure 54, undoubtedly proving the substitution pattern.

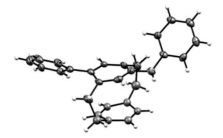

Figure 54. Molecular structure of *para*-phenylated *N*-phenylamino[2.2]paracyclophane **132a** drawn at 50% probability level.

The molecular structure of **133** additionally shows the anticipated C–N cross-coupling, but only after the *para*-arylation took place.

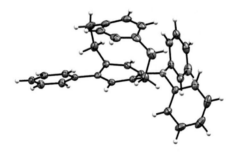

Figure 55. Molecular structure of *para*-phenylated *N,N*-diphenylamino[2.2]paracyclophane **133** drawn at 50% probability level.

To further determine the preference between C–C and C–N cross-coupling a temperature screening with TLC monitoring was conducted in order to verify, that only the C–N arylation proceeds above 80 °C, while the C–C arylation already yields **132a** at 40 °C.[139]

Since the *para* C–C activation is the more intriguing reaction and is another example of the numerous peculiarities reported for [2.2]paracyclophane, all further investigations focused on this C–C cross-coupling. Since the chromatographic properties of **36** and **132a** were nearly identical, purification proved to be complicated. Therefore, calibrated gas chromatography with an FID detector was used to further optimize and investigate the reaction.

3.3.2   Optimization of Conversion and Selectivity

The initial screening focused on temperature. Additionally, three halobenzenes were tested to get an insight on the halide preferences (Table 15). The reaction time was fixed to 16 h. Since the side product **133** was too heavy to be detected *via* gas chromatography, all screenings focused on the C–C coupled product. As shown in entry 1, all three halides gave **132a** in moderate conversions, due to formation of the **133** side product (detected *via* TLC) obtained from **132a** as an intermediate. The differences between chloro, bromo, and iodobenzene was insignificant. Upon lowering the reaction temperature to 100 °C (entry 2), the conversion slightly improved. Upon further decreasing the temperature to 80 °C (entry 3), the conversion for chlorobenzene reaches a moderate conversion of 60%, while the conversions for the bromo- and iodobenzene stay the same. At 60 °C, the conversions significantly drop to the 21–31% range. Monitoring of TLC showed that side product **133** is no longer formed at this temperature. Lowering the temperature to 40 °C (entry 5) and room temperature (entry 6), the conversion falls off to the 11–16% range. This first temperature screening was conducted in the context of a supervised Bachelor thesis.[139]

Table 15. Screening of temperature dependence and halide on the conversion to **132a**. All reactions were performed on a 250 µmol scale with biphenyl as internal standard and the conversions were determined by gas chromatography.[139]

Entry	Temperature [°C]	Halide	Conversion [%]	Halide	Conversion [%]	Halide	Conversion [%]
1	120	PhCl	43	PhBr	45	PhI	37
2	100	PhCl	40	PhBr	47	PhI	48
3	80	PhCl	60	PhBr	46	PhI	49
4	60	PhCl	21	PhBr	30	PhI	31
5	40	PhCl	16	PhBr	14	PhI	15
6	r.t.	PhCl	11	PhBr	11	PhI	11

Next, the catalyst was screened as shown in Table 16. The temperature was fixed to 60 °C so that the formation of **133** could be excluded. This was accompanied by an increase of the equivalents of the bromobenzene to three. While the initial catalyst system Pd_2dba_3 / SPhos gave a nearly identical conversion as during the temperature screening with 31% (entry 1), the tetrakis(triphenylphosphine)-palladium(0) catalyst in entry 2 gave a conversion of 52%. All other tested catalysts, such as $Pd(OAc)_2$, $[Pd(dppf)]Cl_2$, $[Pd(PPh_3)]Cl_2$ and PEPPSI-IPr (entries 3–6) gave no conversion. From this screening it can be concluded, that Pd(0) catalysts do promote the C–H activation, while all Pd(II) catalysts are inactive in both C–H activation and C–N cross-coupling, which was monitored by TLC.

Table 16. Screening of catalyst systems. All reactions were performed on a 250 µmol scale with biphenyl as an internal standard and the conversions were determined by gas chromatography.

PhBr (3.0 equiv.)
[Pd] (5 mol%)
KO^{t}Bu (2.0 equiv.)
toluene
60 °C, 16 h

36 → **132a**

Entry	Catalyst (5 mol%)	Conversion [%]
1	Pd$_2$dba$_3$ / SPhos (15 mol%)	31
2	Pd(PPh$_3$)$_4$	52
3	Pd(OAc)$_2$	0
4	[Pd(dppf)]Cl$_2$	0
5	[Pd(PPh$_3$)$_2$]Cl$_2$	0
6	PEPPSI-IPr	0

Next, the influence of monodentate phosphine ligands was evaluated (Table 17) with Pd$_2$dba$_3$ as palladium source. Both, the already tested SPhos (entry 1) and XPhos (entry 2), perform with identical conversions of 31%, while RuPhos gives a lower conversion of 22% and ^{t}BuPhos is nearly inactive with only 2%. In summary, the Pd(PPh$_3$)$_4$ catalyst system (Table 16, entry 2) is superior to all of the tested Pd$_2$dba$_3$/ligand systems.

Table 17. Screening of phosphine ligands. All reactions were performed on a 250 µmol scale with biphenyl as an internal standard and the conversions were determined by gas chromatography.

PhBr (3.0 equiv.)
Pd$_2$dba$_3$ (5 mol%)
ligand
KO^{t}Bu (2.0 equiv.)
toluene
60 °C, 16 h

36 → **132a**

Entry	Ligand (15 mol%)	Conversion [%]
1	SPhos	31
2	XPhos	31
3	RuPhos	22
4	^{t}BuPhos	2

The next parameter to screen was the influence of solvent as summarized in Table 18. Again, the initial solvent toluene was superior to all other polar solvents examined.

Table 18. Screening of solvents. All reactions were performed on a 250 μmol scale with biphenyl as an internal standard and the conversions were determined by gas chromatography.

Entry	Solvent	Conversion [%]
1	Toluene	46
2	THF	3
3	DMF	4
4	Dioxane	4

The influence of base was evaluated next (Table 19). Once more, the initially used KO^{t}Bu (entry 1) was the best base with 51% conversion, while Cs$_2$CO$_3$ gave only trace amounts. KOH and K$_3$PO$_4$ were inactive in promoting the C–H activation.

Table 19. Screening of bases. All reactions were performed on a 250 μmol scale with biphenyl as an internal standard and the conversions were determined by gas chromatography.

Entry	Base	Conversion [%]
1	KO^{t}Bu	51
2	Cs$_2$CO$_3$	2
3	KOH	0
4	K$_3$PO$_4$	0

Finally, the influence of equivalents of KO^{t}Bu were screened as a last variable to influence the conversion. While an excess of base with 8 and 4 equivalents (entries 1–2) impede the reaction with conversions of 15% and 32%, respectively, the optimal number of equivalents is slightly above equimolar amounts (entries 3–5) giving conversions of 51–46%. Equimolar and sub-equimolar amounts of KO^{t}Bu (entries 6–7) again hamper the reaction with 29% and 23% respectively. The absence of base (entry 8) prohibits the reaction completely. Therefore, a decrease of the base to 1.25 equivalents was chosen in in further investigation.

Table 20. Screening of equivalents of KO^{t}Bu. All reactions were performed on a 250 μmol scale with biphenyl as an internal standard and the conversions were determined by gas chromatography.

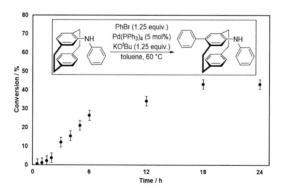

Entry	Equivalents of KO^{t}Bu	Conversion [%]
1	8.0	15
2	4.0	32
3	2.0	51
4	1.5	47
5	1.25	46
6	1.0	29
7	0.75	23
8	0	0

A time profile of the reaction with aliquots taken out of the reaction in defined intervals (Figure 56) with reduced bromobenzene and base equivalents was recorded. Interestingly, the reaction takes a while to increase the conversion rate with the 10% conversion threshold reached after 2 h. After 18 h, the conversion rate drops, staying in the 40–50% range.

Figure 56. Time profile of the formation of **132a**. All values are calculated from an average of two reactions. All reactions were performed on a 250 μmol scale with biphenyl as internal standard and the conversions were determined by gas chromatography. Error bars are set to the highest measured standard deviation of 2.6%

3.3.3   Scope of Derivatives

The optimized reaction conditions for **133** were determined to be Pd(PPh₃)₄ (5 mol%), KOᵗBu (1.25 equiv.) in toluene at 60 °C for 24 h. The amount of bromoarene reactant was set to 1.25 equivalents in order to prevent subsequent C–N cross-coupling similar to **133**. With these reaction conditions, numerous aromatic bromides were screened (Table 21). Electron poor bromoarenes (entries 1–5) such as 4-nitro, 4-nitrile, 4-trifluoromethyl or 4-acetyl did not give the desired products. Unfortunately, the 4-triazinyl phenyl did not give any product either, despite that it would have yielded another interesting TADF structure (compare chapter 3.2). Next, electron donating bromo arenes were tested (entries 6–11), which yielded the desired respective products in varying yields. In each case the target structure was either verified by single crystal analysis or the presence of the N–H band in IR spectroscopy.

Table 21. Summary of the tested scope of derivatives.

Entry	R-Br	Code	Isolated yield [%]
1	O₂N–⬡–Br	b	–
2	NC–⬡–Br	c	–
3	F₃C–⬡–Br	d	–
4	Me(O)C–⬡–Br	e	–
5	Ph,N=triazine–⬡–Br	f	–
6	anthracenyl–Br	g	57
7	Ph₂N–⬡–Br	h	43
8	carbazolyl–⬡–Br	i	12

9	Me —Br Me	j	82
10	MeO——Br	k	39
11	—Br OMe	l	44
12	N——Br	m	–
13	N —Br	n	–
14	N —Br N	o	–
15	Ph N N——Br N Ph	p	–
16	O —Br	q	–

While the 2-anthracenyl derivative **132g** and the diphenylamino derivative **132h** were isolated in moderate yields, respectively (entries 6, 7), the analogous carbazolyl derivative **132i** could only be isolated in a poor yield of 12% (entry 9). Despite having an *ortho*-methyl group, which induces steric hinderance and should therefore hamper the performance of the reaction, the *para*-xylene derivative **132j** was formed with the best yield in this screening with 82% (entry 9). To corroborate the marginal influence sterics seem to have on the yield of the reaction, *para*-methoxy (39%, entry 10) and *ortho*-methoxy (44%, entry 11) derivatives were synthesized. For both examples, the absolute configuration could be determined by crystallography (Figure 57 for **132k** and Figure 58 for **132l**).

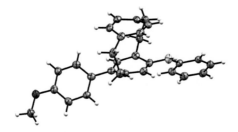

Figure 57. Molecular structure of **132k** drawn at 50% probability level.

Even though the yield deteriorated when the methoxy group was directly adjacent to the coupling position, it did not significantly influence the yield. To conclude, heteroaromatic bromides were tested and neither the electron poor derivatives such as pyridyls (entries 12, 13), pyrimidyl (entry 14) or triazine (entry 15), nor the electron rich furanyl (entry 16) yielded any product.

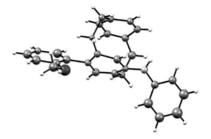

Figure 58. Molecular structure of **132l**. Due to insufficient refinement only the connectivity was verified.

In summary, the tested reaction gives access only to phenyl and electron rich arenes. Despite its drawbacks, this method is suitable to construct bulky steric environments, as *ortho*-substituents such as methyl and methoxy can be successfully without reducing yields significantly.

3.3.4   Proposed Mechanism

Finally, an insight into the mechanism will be discussed to ascertain how and why the C–H activation is targeting the *para*-position in the [2.2]paracyclophane scaffold. The proposed mechanism is depicted in Scheme 72. All cross-coupling reactions begin with an oxidative addition of the haloarene to the metal center, the key question then is how the palladium complex **134** interacts with the the [2.2]paracyclophane-bridged diphenylamine **36**. Since the reaction only proceeds in the presence of base (see Table 20), it is reasonable to assume, that **36** is participating as the conjugate base. The oxidative addition complex **134** is electrophilic and therefore attacks the most electron dense position. In the case of **36**, this is the [2.2]paracyclophane scaffold as additional electron density is present from the adjacent second deck. The η^6-palladium(II) complex **135** is expected to form. Similar complexes are reported to form from [2.2]paracyclophane with ruthenium(II)[162], chromium(III)[163] or tripalladium(II).[164] This pronounced ligating property of [2.2]paracyclophane is caused by the reduced antibonding inter-deck overlap and is thoroughly asserted by Rozenberg *et al.*[165] The π-complex must then collapse selectively to the σ-complex **136** to get a *para*-functionalization. This is supported by a quinoimine intermediate, which have been reported in [2.2]paracyclophane chemistry.[166] Related products obtained from oxidized intermediates, comparable to the here hypothesized intermediate **136** were reported by Bolm *et al.*,[40] strongly supporting the here proposed mechanism. In the final step, a reductive elimination procceds to yield **132** and the palladium(0) species.

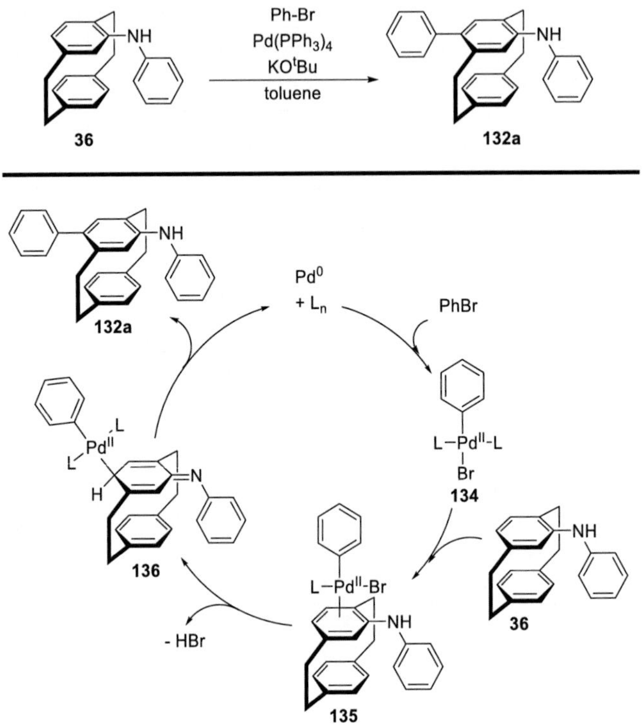

Scheme 72. Proposed mechanism of C–H activation on N-phenylamino[2.2]paracyclophane **36**.

In summary, a protocol was developed using a [2.2]paracyclophane-bridged diphenylamine **36**, which undergoes *para*-C–H activation as the preferred reaction pathway contrary to the expected Hartwig-Buchwald cross-coupling. It was also found that this reaction proceeds in the presence of KO^{t}Bu and palladium(0) catalyst precursors in toluene with electron-rich haloarenes at rather mild reaction conditions. Bulky substituents are tolerated and the mechanism proposed takes into account the peculiarity of [2.2]paracyclophane to I) form η^6-complexes and II) the formation of a quinoimine derivative **136**. Both assumptions are strongly supported by literature (compare chapter 1.1.5).[40, 165-166]

3.4   Chiral Chemical Vapor Deposition Polymers

Lahann *et al.*[69] recently expanded the scope of the CVD process to directed polymerization of surface-anchored nanofibers templated (see chapter 1.2) by the properties of the liquid crystals (LC). By using cholesteric (chiral nematic) liquid crystal phases, a chiral templating effect could be demonstrated in the form of a wide spiral twisting of the whole fiber. In other words, a top-down template for nanoscale chirality of molecularly nonchiral fibers was demonstrated for CVD. This gives rise to the inverse question if within a nonchiral nematic (homeotropic) phase a chiral monomer would be able to induce the same chiral information to obtain helical fibers and whether the twisting angle and spiral width is comparable (Scheme 73). This will only be possible if the quaternary stereogenic center at the alcohol is stable under the monomer pyrolysis conditions.

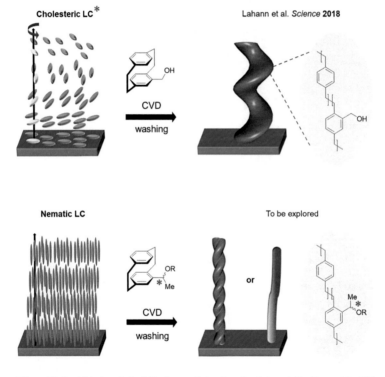

Scheme 73. Established method of chirality templating from the cholesteric liquid crystal (top)[69] and the now to be explored question of chirality transfer from a chiral CVD monomer (bottom). Chiral information indicated with asterisk.

In order to tackle these questions, a synthetic access to sufficiently enantioenriched chiral precursors was needed to provide the Lahann group *via* a collaboration with suitable CVD monomers. A methyl group was chosen to introduce chirality as this is easily available from the 4-acetyl[2.2]paracyclohane (**137**) *via* reduction. As substituted [2.2]paracyclophanes exhibit planar chirality, a nonselective reduction yields a complex mixture of 4 different molecules as indicated in Scheme 74. Most notably, a set of enantiomers will yield a racemic parylene (**PPX**) polymer, while a set of enantiopure diastereomeric alcohols will yield enantiopure parylenes if the quaternary chiral center is stable to racemization at the CVD furnace temperature and contact time.

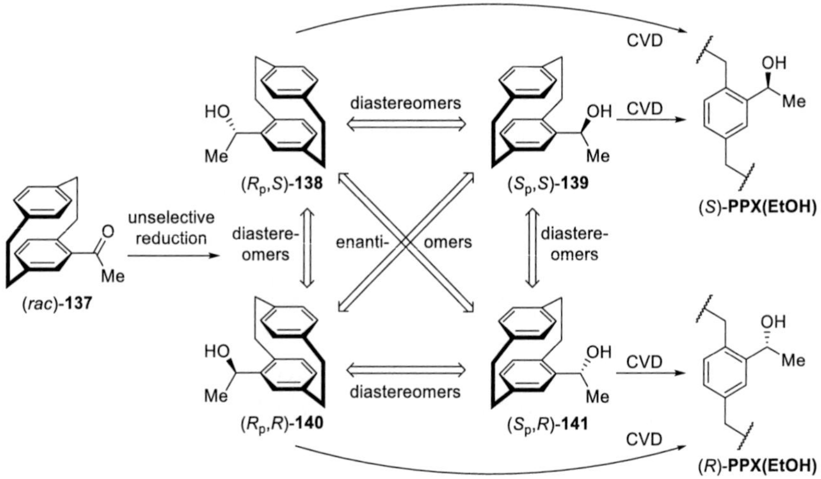

Scheme 74. Overview of potentially available chiral CVD presursors from reduction of **137** and the chirality of the resulting parylene (**PPX**).

Since diastereomers can easily be separated by chromatography, but enantiomers need advanced purification techniques, an asymmetric reduction protocol was chosen.

3.4.1   Preparation of Chiral Precursors *via* Corey-Itsuno-Reduction

This method of asymmetric reduction (also known as CBS reduction) of prochiral aromatic ketones was first described by Itsuno *et al.* in 1983[167] by using chiral 1,2-aminoalcohols for a templated BH$_3$ reduction.[168-169] These 1,2-aminoalcohols can be easily obtained from the chiral pool *via* reducing α-amino acids. Soon thereafter Corey, Bakshi, Shibata *et al.* (hence CBS) further investigated the mechanism, both isolating an catalytically active oxaborozolidine species and compiling a set of guidelines for selectivity prediction.[170] The diphenylated boroxazolidine species derived from proline[170] and especially B-functionalized derivatives such as (*S/R*)-**142** were established as a stable, well-performing and now commercially available auxiliary and catalyst for asymmetric reduction of ketones. This protocol quickly gained wide application in medicinal and natural product chemistry.[171] A thorough evaluation of the mechanism as shown in Scheme 75 for the (*S*)-CBS catalyst visualizes that the steric congestion for the catalytic key intermediate for the synthesis of CVD precursors is significantly different for the planar enantiomers, as the preorganization by coordination of the ketone to the boron favors an anti-configuration of the [2.2]paracyclophane relative to the azolidine moiety. When the second "deck" of the [2.2]paracyclophane points towards the boronhydride species, an adverse reduction efficiency can be expected as the intramolecular hydride transfer is disfavored by an enlarged transfer distance. This is in accordance with the results that (S_p)-[2.2]paracyclophane structure is less favored for reduction by the (S)-CBS catalyst.[172] Additionally with equimolar amounts of BH₃, the nucleophilic activation by the boron coordination to the ketone makes the complex prone to reduction by a noncoordinated boron hydride species from the reaction mixture. With regard to this thesis the synthetic approach to the chiral CVD precursors was first conducted as described for **137** by Kagan *et al.* at 0 °C in dry tetrahydrofurane,[172] and the $(S_p,S)/(R_p/R)$ pair of enantiomers was isolated and the *ee* was measured by chiral HPLC.

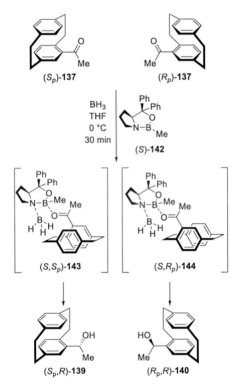

Scheme 75. Corey-Itsuno reduction of (rac)-4-acetyl[2.2]paracyclophane shown for the CBS catalyst (S)-**142** with the key intermediate shown for each enantiomer.

The acetyl[2.2]paracyclophane **137** was synthesized using acyl chloride under aluminium chloride Lewis acidic conditions by a known procedure (Scheme 76) in 55% yield.[142]

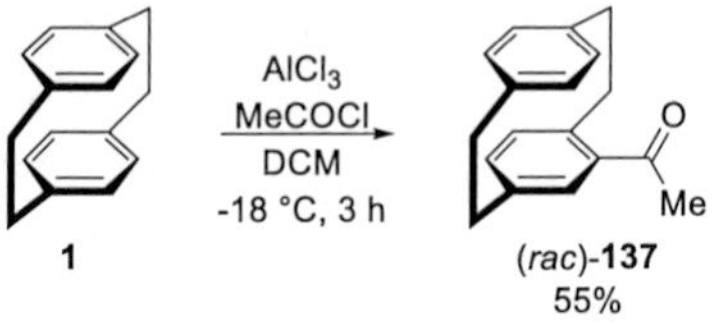

Scheme 76. Synthesis of (*rac*)-4-acetyl[2.2]paracyclophane **137**.

As shown in Table 22, none of the tested reduction conditions were able to yield sufficiently enantiopure alcohols. Even at stoichiometric loadings of CBS catalyst (entries 5–7), an enantiomeric excess of >95% could not be obtained. This can be explained by residual $NaBH_4$ species in the commercially obtained CBS catalyst solutions[173] or the previously discussed increased steric congestion impeding an efficient coordination to the CBS catalyst for both starting material enantiomers. The enantiomeric excesses reported by Kagan *et al.* with a 15% catalyst loading at 0 °C was significantly depended on the equivalents of BH_3 used, with 0.30 equiv. yielding 90%ee. This could be improved by using $BH_3 \cdot Me_2S$ adducts and a solubilized addition of the ketone. These variations were not further investigated, as another approach (discussed below) was yielding better results.

Table 22. Overview of tested Corey-Itsuno reduction conditions varied from Kagan *et al.*[172]

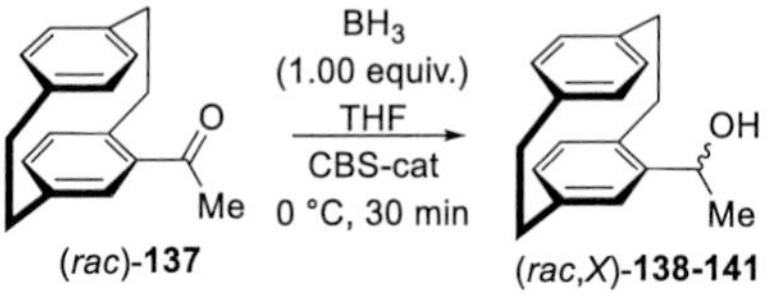

Entry	Catalyst	Conditions	ee[a] of S_p/S	ee[a] of R_p/R	Conversion
1	10% (*R*) cat	RT, 30 min	38%	–	quant.
2	10% (*S*) cat	RT, 30 min	–	40%	quant.
3	10% (*R*) cat	RT, 30 min	72%	–	quant.
4	20% (S) cat	RT, 30 min	–	44%	quant.
5	100% (*S*) cat	0 °C, 30 min	–	72%	quant.
6	100% (*S*) cat	0 °C, 30 min	–	74%	quant.
7	100% (*R*) cat	0 °C, 30 min	80%	–	quant.

BH_3 solution and the respective CBS catalyst were mixed and stirred for 5 min before addition of the ketone as a solid. – a: enantiomeric excess measured by chiral HPLC.

3.4.2   Chiral Resolution *via* Derivatization

Another route to obtain enantio- and diastereomerically pure alcohols as CVD precursors is to first prepare an enantiomerically pure [2.2]paracyclophane carbonyl compound. The advantage of this route is that the reduction can be performed in a nonselective manner, as the resulting diastereomeric mixture is enantiomerically pure. For that, the acetyl[2.2]paracyclophane **137** was substituted with the formyl[2.2]paracyclophane **5**, as a convenient procedure for the enantiopure separation by chiral resolution is reported in literature.[48] The racemic aldehyde **5** was prepared by Rieche formylation (see Scheme 38). In the next step, racemic **5** was condensated with chiral amine **47** in boiling toluene overnight to yield a diastereomeric mixture of (S_P,R)-**48** and (S_P,S)-**48**. Repeated recrystallization in *n*-hexane (up to 6 iterations), yields the (S_p,R)-**48** diastereomer as a pure compound in the form of transparent cubic crystals in 26% yield based on 50% of the available (S_p) enantiomer in the racemic mixture of **5**.

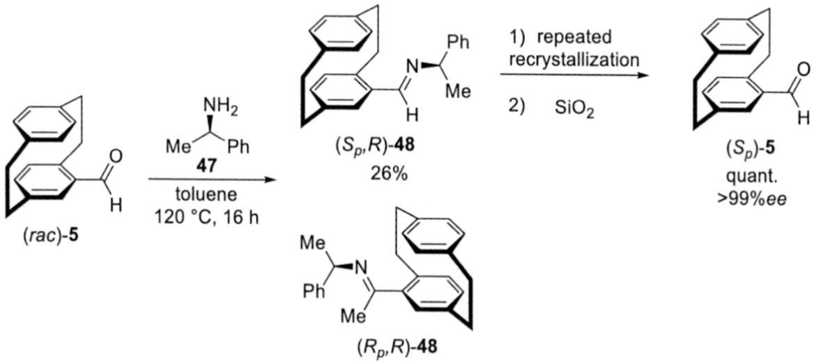

Scheme 77. Chiral resolution of (*rac*)-4-formyl[2.2]paracyclophane (**5**) *via* diastereoselective recrystallization of the respective (*R*)-imine condensate (S_p,R)-**48**.[48]

The %*de* can easily be monitored by ¹H NMR spectroscopy *via* the imine proton singlet at 8.38 ppm and the methyl doublet at 1.68 ppm as shown in Figure 59.

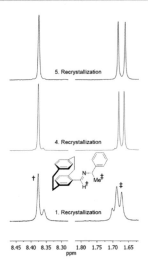

Figure 59. Set of ^{1}H NMR cuts of the diastereomerically relevant signal ranges for the diastereomeric mixture of **48** with increasiny diastereopurity.

Subsequent stirring of (S_p,R)-**48** in dichloromethane with silica gel and filtering yields quantitative amounts of the (S_p)-aldehyde **5**. This aldehyde was then alkylated with methyllithium to yield a diastereomeric mixture of enantiomerically pure alcohols (Scheme 78). Although the significant discrepancy in yield can be explained by the tedious chromatographic workup, an influence of the steric bulk on the trajectory of the nucleophilic attack of methyllithium is to be taken into account. A molecular analysis showed that the relative position of the aldehyde to the [2.2]paracyclophane is reported as showing towards the ethylene bridge and as showing towards the *ortho*-sp^2-proton for comparable diformyl derivatives.[174] It appears that the aldehyde moiety can freely rotate, putting aside any steric arguments for the discrepancy in yield.

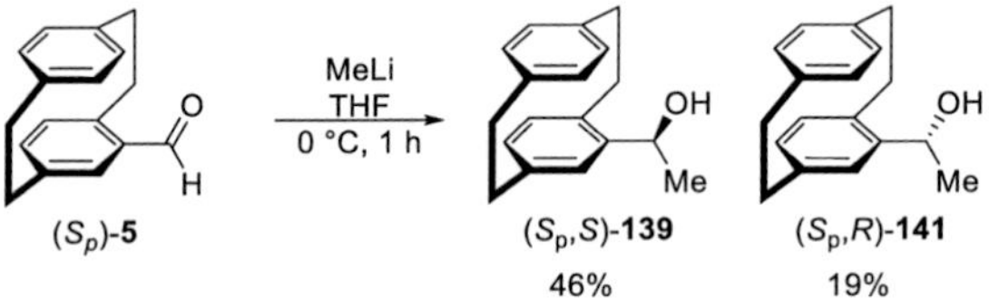

Scheme 78. Unselective methylation of enantiopure aldehyde **5**.

To determine the absolute configuration a thorough analysis was conducted for the two diastereomers. A crystal structure could be obtained for (S_p,R)-**141** giving access to the absolute configuration based on the known stereochemistry of the [2.2]paracyclophane moiety.

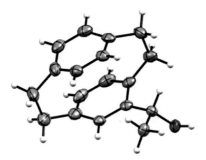

Figure 60. Molecular structure of (S_p,R)-**141** drawn at 50% probability level.

In contrast, no single crystal analysis of the (S_p,S)-**139** diastereomer could be obtained, therefore NOESY experiments were conducted to assign the absolute configuration to the quaternary stereogenic center of the alcohol. Three relevant hydrogens were identified on the [2.2]paracyclophane structure and their 1D slices were extracted to compare their interaction to the groups directly located to the quaternary stereocenter (Figure 61). The NOE signal intensity was interpreted as distance between the hydrogen atoms with a decreasing signal for increasingly remote atoms.[175] The pseudo-*geminal* hydrogen, marked in green shows equally strong NOE signals with both the hydroxyl and the methinyl groups, while no NOE signal is detected for the methyl group. This indicates that the methyl group points away from the second phane deck. The *ortho* hydrogen marked in red shows a predominant NOE interaction with the hydroxyl group, indicating that it is facing away from the *ortho*-ethylene bridge of the [2.2]paracyclophane. This is verified by the NOE slice of the adjacent aliphatic hydrogen on the ethylene bridge marked in blue. It shows significant NOE interaction with the methyl and methinyl groups. In summary an absolute configuration can be assigned to the quaternary stereocenter as (S) knowing that the substituted [2.2]paracyclophane **5** possessed (S_p) configuration.

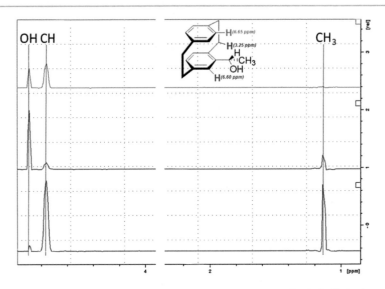

Figure 61. Relevant slices (green, red and blue) of the NOESY experiment for (S_p,S)-**139** showing their through space interaction with the groups attached to the chiral quaternary center. The section between 2–4 ppm was omitted to remove the very intense diagonal peak of the aliphatic hydrogen.

From a stereochemical point of view it is noteworthy, that in both diastereomers the methinyl hydrogen is directed towards the sterically most congested direction of the second phane deck and the ethyl bridge, respectively.

3.4.3   CVD Polymerization of Chiral Alcohols

These two compounds were provided to Kenneth Cheng at the Lahann group to conduct LC-templated CVD polymerization experiments. The results of this collaboration are presented in this subchapter.[105] Initial results showed that the nanofibers obtained from the chiral precursor (S_p,S)-**139** indeed show I) helicity and II) all individual fibers show the same clockwise (c.w.) sense of rotation, as shown in the SEM image in Scheme 79. In contrast to a cholesteric liquid crystal template, the fibers exhibit a non-bent rod-like shape with a comparably tight twisting along the fiber axis. A chemical surface analysis (XPS and IRRAS) of the fibers was conducted to demonstrate that the chemical composition is similar to the precursor and a non-templated parylene film, and that the hydroxyl group is therefore still present in the fibers. To further evaluate the influence of molecular chirality on nanofiber formation in an achiral nematic liquid crystal as a template, both enantiomers were polymerized and a mirror-inverted behavior was observed as shown in Figure 62 with the fibers obtained from (S_p,S)-**139** all show clockwise (c.w.) helicity, while the fibers obtained from (S_p,R)-**141** all show counter-clockwise (c.c.w.) helicity.

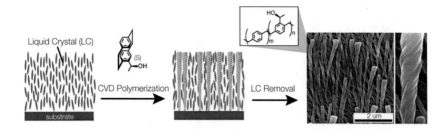

Scheme 79. Formation process in MDA-98-1602 liquid crystal and SEM image of helically twisted nanofibers obtained from the chiral precursor **139**. Results and images provided by Dr. Cheng.[105]

An analysis by circular dichroism (CD) spectroscopy of the dispersed nanofibers additionally verify by the mirrored CD signal of the $\pi \rightarrow \pi^*$ transition at 250–300 nm that the diastereomers provided to form the helical nanofibers behave as enantiomers after the CVD process.

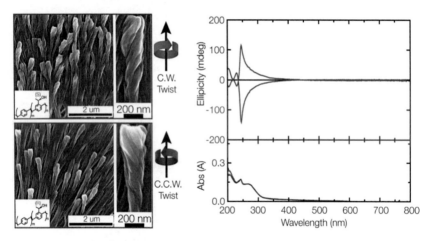

Figure 62. SEM images of the mirror-inverted helical nanofibers prepared from the chiral precursors **139** and **141** and the corresponding CD spectra of the dispersed helical nanofibers. Results and images provided by Dr. Cheng.[105]

Therefore, the initial hypothesis of chirality transfer from a molecule to the micrometer scale was proven. To further explore the limits of chirality transfer, a screening of the influence of enantiomeric excess (%*ee*) was conducted as shown in Figure 63 by mixing of the provided enantiomerically pure diastereomers **139** and **141**. It was observed, that the twisting angle is directly dependent on the amount of enantiomeric excess of the starting material.

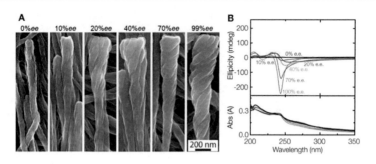

Figure 63. Influence of enantiomeric excess on fiber helicity shown as SEM images (A) and the respective CD signals (B) of dispersed fibers. Results and images provided by Dr. Cheng.[105]

Finally, a combination of the precursor-based helicity templating effect and the cholesteric liquid crystal based templating effect (see chapter 1.2) was conducted to observe either a cooperative or a competitive influence on fiber formation. While the cholesteric liquid crystal provides a comparatively wide spiral of the whole fiber,[69] the precursor templating yields a linear fiber with a tight twist along the fiber axis. When combining a chiral liquid crystal and chiral precursor, both inducing a clockwise helicity as shown in Figure 64 top, both templating effects are cooperating to yield a fiber which is both helical along the fiber axis and helical in the form of a spiral in the micrometer range. In contrast (Figure 64 bottom), when combining a chiral liquid crystal and chiral precursor, who are both known to induce helicity in opposite senses of rotation, the resulting fiber still exhibits the helical twist along the fiber axis caused by the chiral precursor, while the wider spiral helicity from the cholesteric liquid crystal is no longer observed.

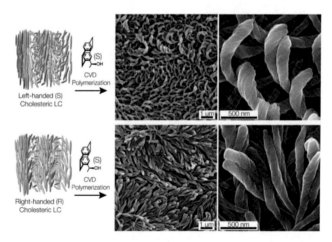

Figure 64. Cholesteric LC templated CVD of chiral precursors **139** and **141** with respective SEM images of the fibers. Results and images provided by Dr. Cheng.[105]

3.4.4   Synthesis of Chiral Methyl Ethers

Encouraged by these promising results, the question arose whether the chiral induction from the monomer is predominantly caused by hydrogen bond interaction, of simply by steric effects. The role of hydrogen bonding had to be assessed as this property is commonly utilized for self-assembly of helical polymers and supramolecular architectures.[176-180] For this purpose, precursors **139** and **141** were methylated using methyliodide under basic conditions (Scheme 80).

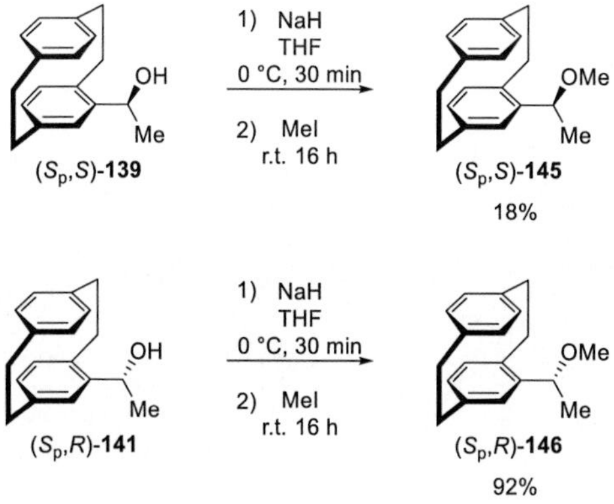

Scheme 80. Methylation of CVD precursors **139** and **141** to access nonacidic CVD precursors **145** and **146**.

A single crystal was obtained for (S_p,R)-**146** with the absolute configuration assigned based on the known stereochemistry of the [2.2]paracyclophane moiety (Figure 65).

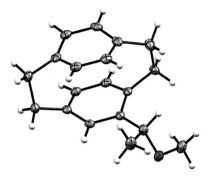

Figure 65. Molecular structure of (S_p,R)-**146** drawn at 50% probability.

3.4.5   CVD Polymerization of Chiral Ethers

These two compounds were provided to Kenneth Cheng at the Lahann group to conduct LC-templated CVD polymerization experiments. The results of this collaboration are presented in this subchapter.[105] Interestingly, upon templated CVD polymerization much more brittle and less tightly twister fibers were obtained which still presented an exclusive helical sense of rotation (Figure 66). The senses of rotation were identical to the ones observed for the free alcohol precursors. Ultimately indicating that hydrogen bonding may play less of a role on chiral induction, but is significant in order to obtain stout fibers. Bigger substituents were not successfully applied as heavier precursors tend not to outlast the CVD process.

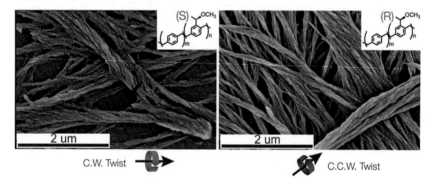

Figure 66. SEM images of CVD fibers from methyl protected chiral precursors **145** and **146**. Results and images provided by Dr. Cheng.[105]

3.4.6   Synthesis of Chiral Fluorinated Alcohols

Another drawback of the liquid crystal templated CVD polymerization was the availability of the liquid crystals, for which the commercial sources often did not disclose the chemical structure, but it was known to contain fluorinated groups. For this, an access to fluorinated alcohols, preferably in an enantio- and diastereomerically pure form was needed. Therefore electrophilic trifluoracetylation was conducted on [2.2]paracyclophane **1** under aluminium chloride Lewis acidic conditions using trifluoracetic anhydride (TFAA)[181] to give the desired (*rac*)-4-trifluoroacetyl[2.2]paracyclophane (**147**) in 69% yield.

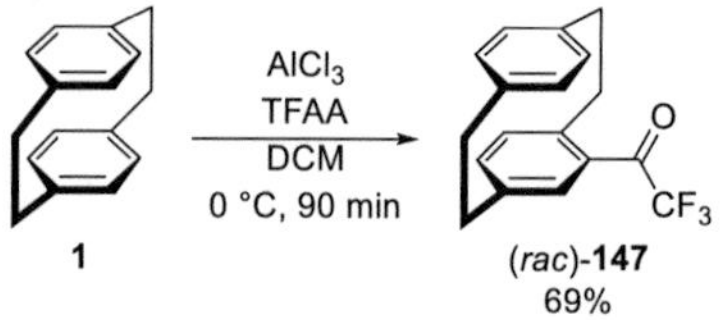

Scheme 81. Synthesis of (*rac*)-4-trifluoroacetyl[2.2]paracyclophane (**147**).

To evaluate the purification process of an enantioenriched and diastereomerically pure trifluoromethyl alcohol, a racemic and diastereomeric mixture was synthesized. Therefore, the racemic ketone **147** was reduced using LiAlH$_4$ to yield the trifluoromethyl methanol **148** as a diastereomeric mixture in 93% yield (Scheme 82). Interestingly, the resulting diastereomeric mixture was inseparable already under analytical conditions such as TLC and analytical HPLC. This is predominantly caused by the lipophilicity, acidity and hydrogen bonding effects[182] of the trifluormethyl moiety.

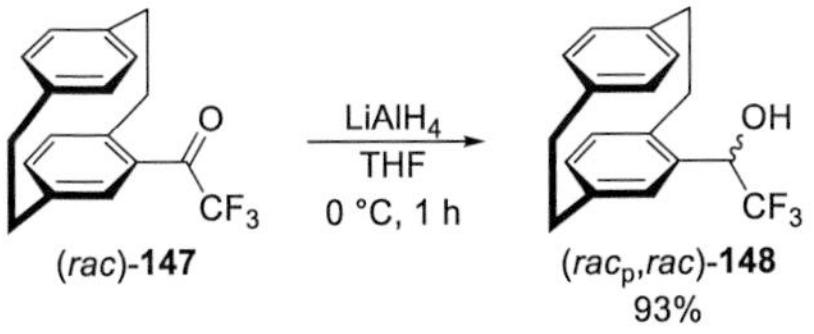

Scheme 82. Racemic reduction of ketone **147** to the chiral alcohol **148** using LiAlH$_4$.

Taking into account the previously discussed structural differences of the nonfluorinated alcohols (**139** and **141**), where in both cases the methinyl group of the alcohols were directed towards the pseudo-*geminal* hydrogen and the methyl and hydroxyl group switched positions for that stereogenic center, the increased polarity of the trifluoromethyl group might be an explanation for the minute disparity of the retention factor. In other words, the resulting dipole moment of a methyl and hydroxyl group at a tetrahedral carbon center is very much aligned along the C–O bond, while for a trifluoromethyl and a hydroxyl group at a tetrahedral carbon center, the dipole moment is more averaged between these two polar groups, diminishing the difference of the chromatographic properties for a set of diastereomers.

3.4.7   Gas-phase Hydrogen Interaction Monomers

In another project funded by the SFB1176, project B3, interaction *via* hydrogen bonding was to be studied in order to access further tools to control the properties of parylenes. In principle this is possible by hydrogen bond interaction in the gas phase (**150**) or on surface before polymerization, or as a chemical feature of the resulting parylene (**PPX(CONH₂)**) as depicted in Scheme 83. As it is known that the free carboxylic acid of [2.2]paracyclophane is unstable, yet their respective ketones are suitable for CVD conditions,[58] an evaluation of the more stable amides such as **151** and the respective secondary derivatives was to be conducted. In this regard, a suitable synthetic approach was taken.

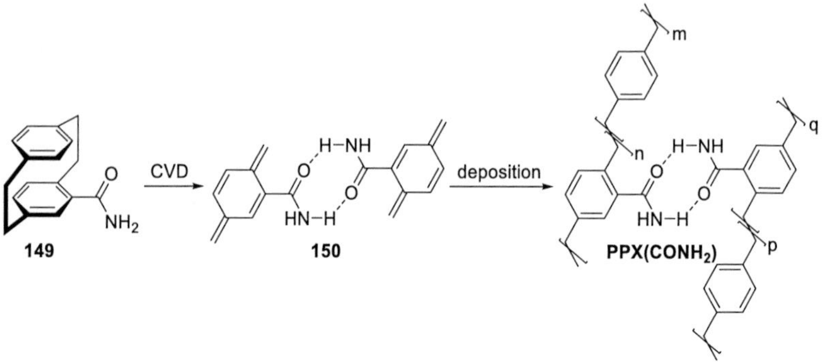

Scheme 83. Concept of hydrogen interaction for the formation of new parylenes.

As no stereogenic center remains after CVD, all syntheses were performed using racemic compounds. In order to obtain the primary amide **151**, the nitrile **97** was chosen as an important intermediate. In contrast to known methods starting from the 4-bromo[2.2]paracyclophane which is converted to nitrile **96** using an excess of CuCN,[126, 183] or of a carboxylic acid chloride intermediate,[184] the previously developed method of aldoxime intermediate **96** formation with subsequent dehydration under one-pot conditions was used to obtain nitrile **97**. Then a protocol from Katritzky et al.[185] using hydrogen peroxide as oxidant in DMSO under basic conditions was adapted to **97** to obtain the desired primary amide **151** in excellent yield (Scheme 84).

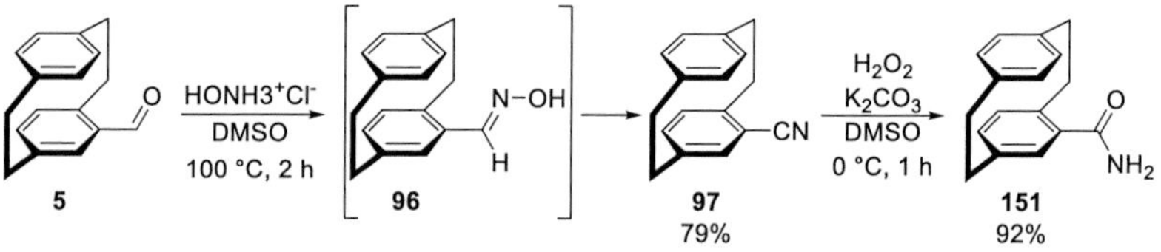

Scheme 84. Synthesis of the primary [2.2]paracyclophanecarboxamide.

In order to increase the scope of available amides, two more *N*-functionalized derivatives were prepared. The *tert*-butyl derivative **152** was synthesized *via* a Ritter reaction by heating nitrile **97** in *tert*-butanol under acidic conditions (Scheme 85).

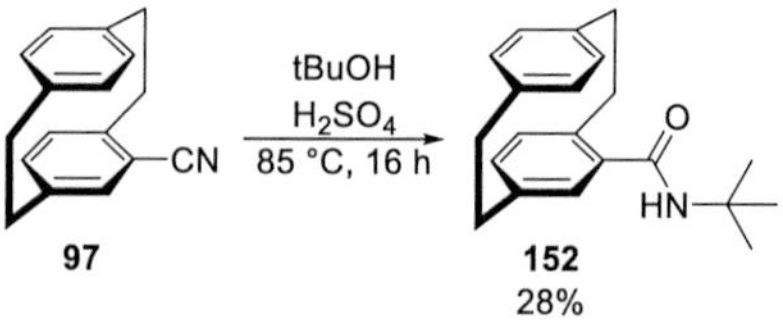

Scheme 85. Synthesis of *N-tert*-butyl-4-[2.2]paracyclophane **152**.

Finally, a *N*-4'-methoxyphenyl derivative **154** was prepared using an acyl chloride route[184] in poor yield (Scheme 86).

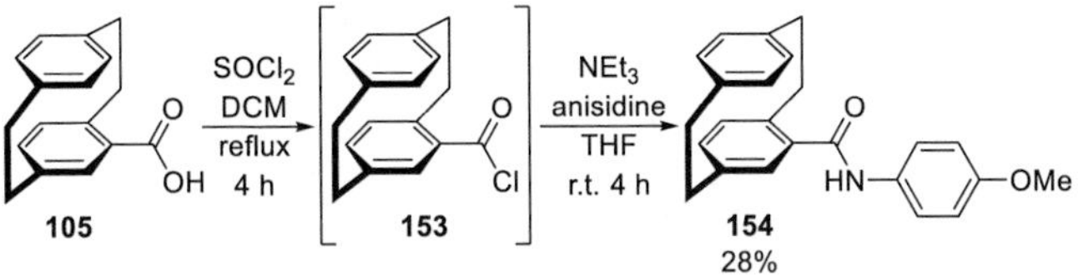

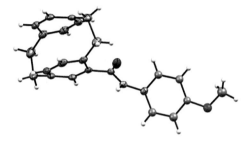

Scheme 86. Synthesis of *N*-4'-methoxyphenyl derivative **154**.

Additionally, a single crystal suitable for crystal analysis was obtained showing a trans configuration of the amide bond (Figure 67).

Figure 67. Molecular structure of carboxamide **154** drawn at 50% probability level.

All synthesized amides were provided to Dr. Hussal at the Lahann group to further investigate their CVD behavior. Unfortunately, none of the synthesized CVD precursors were stable enough and decomposed under CVD conditions, or did not form stable films. Due to these initial results, no further CVD investigation of these compounds was conducted, but a synthetic access to functionalized [2.2]paracyclophanecarboxamides *via* Ritter protocol and a acid chloride protocol was investigated.

In summary, amides proved to be unsuitable substrates for CVD, similarly to the free acid. Therefore, further investigation of the gas-phase hydrogen interaction was stopped.

3.5 Molecular Sensor for Cucurbit[8]uril

In a collaboration with Dr. Stephan Sinn in the Biedermann group at KIT, the 4-pyridyl[2.2]paracyclophane (**157**) was a desired molecule in order to be evaluated as a molecular sensor for a supramolecular indicator displacement assay (IDA) using curcurbit[8]uril (**CB8**) as a supramolecular host. The concept of an IDA is shown in Scheme 87. Hereby a sensitive indicator is pre-complexed in a supramolecular host. If an analyte with a higher binding affinity is present, the indicator is displaced. The gradual displacement of the indicator by the analyte can be followed by a quantifiable spectroscopic change.[186-191]

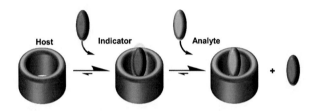

Scheme 87. General schematic representation of an indicator displacement assay (IDA), with the spectroscopically detectable group showing a glow.

It is a great promise of supramolecular chemistry to provide robust, selective and sensitive artificial receptors, that can complement contemporary antibody-based assays in real-world sensing applications,[186, 189, 192-193] such as quick and low-cost monitoring of drug levels. However most synthetic supramolecular sensing methods face great challenges in biofluids such as blood serum because of the presence of numerous possible interferants such as salts, lipids, hormones and proteins that can act as spectroscopic noise,[194] or as competitive binders to the rather unselective artificial hosts. The composition of these biofluids additionally varies for different patients or even situations.[194-197]

Despite many proof-of-principle reports in closely controlled media, only few supramolecular sensing systems are reported to be performants at practical conditions.[197-202] For example, when an Alzheimer's drug, such as memantine is to be monitored, then the therapeutic window lays in the nanomolar to micromolar range (0.03 to 1.00 µM in blood plasma).[203-204] This implies that the binding affinity, K_a, of an artificial receptor for the analyte to be detected in blood serum should reach at least 10^6 M^{-1}. With regard to the construction of an artificial indicator displacement assay (IDA), the macrocyclic cucurbit[n]urils (CBn, n = 5–10)[205-210] are particularly suitable as hosts as they strongly bind a range of bio-relevant organic compounds.[186, 211-217] This group of macrocyclic molecules consists of a defined number of glycoluril monomers linked *via* methylene bridges to form a ring-like structure with an inner cavity of a defined diameter (Figure 68). Therefore, the synthetic strategy to obtain cucurbiturils is *via* condensation of urea with glyoxal.

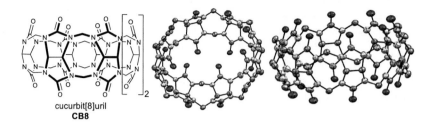

Figure 68. Molecular structure of cucurbit[8]uril (**CB8**) as a representative of this host class. Crystal structure drawn at 50% probability level in two different perspectives, crystal data obtained from CCDC, deposit number 1052403.[218]

As the polar carbonyls show away from the cavity, essentially forming an electron-rich polar portal to the cavity, the cavity itself is less polar. Therefore both neutral and most prominently cationic guests bind to this class of hosts.[208] Common structural units to be bound within the CB*n* hosts include aromatic and aliphatic ammonium compounds, stilbenes or cationically substituted naphthalenes.[208]

In this collaboration, the [2.2]paracyclophane scaffold was simulated to fit within the cavity of cucurbit[8]uril as an *N*-methylated and therefore cationic 4-pyridyl[2.2]paracyclophane **155** (Figure 69).

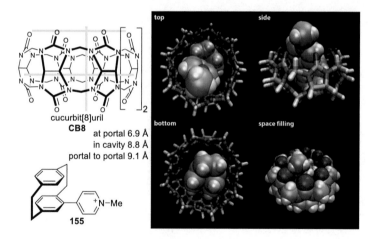

Figure 69. Left: Molecular structures of host **CB8** and guest **152**. Cavity sizes are taken from Kim *et al*.[219] Right: DFT ground-state optimized host:guest complex structure using B3LYP 6-31G(d,p) of **CB8⊃155** in various representation. Results and images were provided by Dr. Sinn.[220]

3.5.1   Evaluation of Reactivity of Suzuki-Miyaura Cross-Coupling[3]

As no synthesis of 4-pyridyl[2.2]paracyclophane **157** is reported in literature, but can be obtained in one step *via* a Suzuki-Miyaura cross-coupling protocol, a catalyst screening was conducted with 4-bromo[2.2]paracyclophane (**17**) as is summarized in Table 23. Although heterocycles are known to be capricious substrates for cross-coupling reactions, the initial protocol using Pd(PPh$_3$)$_2$Cl$_2$ as catalyst with K$_2$CO$_3$ as base in a THF/H$_2$O solvent mixture at reflux (entry 1) already gave 32% yield. A switch of the base to Cs$_2$CO$_3$ gave only a marginal improvement to 35% (entry 2). A change of the catalyst system to Pd$_2$dba$_3$ with SPhos gave a further improved yield of 40%. Further change of the catalyst to Pd(PPh$_3$)$_4$, the base to K$_3$PO$_4$ and the solvent mixture to dioxane/H$_2$O significantly improved the reaction yield to 79%.

Table 23. Screening of Suzuki-Miyaura cross-coupling protocols towards 4-pyridyl[2.2]paracyclophane (**157**).

Entry	Catalyst/Ligand/Base	solvent	Isolated yield [%]
1	Pd(PPh$_3$)$_2$Cl$_2$ (6 mol%) K$_2$CO$_3$ (1.5 equiv.)	THF/H$_2$O	32
2	Pd(PPh$_3$)$_2$Cl$_2$ (6 mol%) Cs$_2$CO$_3$ (1.5 equiv.)	THF/H$_2$O	35
3	Pd$_2$dba$_3$ (3 mol%) SPhos (6 mol%) Cs$_2$CO$_3$ (1.5 equiv.)	THF/H$_2$O	40
4	Pd(PPh$_3$)$_4$ (6 mol%) K$_3$PO$_4$ (1.5 equiv.)	Dioxane/H$_2$O	79

Additionally, the absolute configuration of the target molecule **157**, was determined by single crystal analysis (Figure 70). The angle between the 4-pyridyl substituent and the adjacent deck of the [2.2]paracyclophane scaffold was determined to 43.5°, accompanied by a deck plane distance of 3.08 Å which is nearly identical to the distance reported for the parent scaffold. The longest distance between an ethyl-bridge hydrogen atom and the furthest other aliphatic hydrogen atom is 7.18 Å, while the distance between two aromatic hydrogen atoms is 4.02 Å and 5.50 Å in *para* and in pseudo-*para*

---

[3] Parts of this chapter have been published in:
C. Braun, E. Spuling, N. B. Heine, M. Cakici, M. Nieger, S. Bräse, *Adv. Synth.Catal.* **2016**, *358*, 1664–1670.

position, respectively. Comparing to the reported size of the inner cavity of CB8 (see Figure 69), a smooth fit can be expected judging from the crystal structure as well.

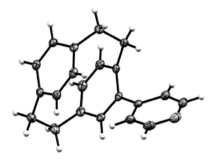

Figure 70. Molecular structure of 4-pyridyl[2.2]paracyclophane (**157**) drawn at 50% probability level.

Thereafter, these reaction conditions were also tested on the centrosymmetric pseudo-*para* dibromide **22** (Table 24). As this molecule displays low solubility in common solvents, it is an even more challenging substrate for cross-coupling reactions. As the substrate possesses two reactive functional groups, the equivalents of base were doubled, but the catalyst loadings were kept identical to the screening in Table 23 to evaluate reduced catalyst loadings.

The initial protocol which gave a poor yield for **157** of 32% (see Table 23, entry 1), gave 30% for **158** (Table 24, entry 1). The doubled amount of reactive bromide groups indicates, that the Suzuki-Miyaurua cross-coupling performs better on the centrosymmetric dibromide **22**. In this reaction, also the mono-coupled side product **159** was isolated in 28%.

Table 24. Screening of Suzuki-Miyaura cross-coupling with the centrosymmetric dibromide **22**.

Entry	Catalyst/Ligand/Base	solvent	Isolated yield of **158** [%]
1	Pd(PPh$_3$)$_2$Cl$_2$ (6 mol%) K$_2$CO$_3$ (3 equiv.)	THF/H$_2$O	30
2	Pd(PPh$_3$)$_2$Cl$_2$ (6 mol%) Cs$_2$CO$_3$ (3equiv.)	THF/H$_2$O	46
3	Pd(PPh$_3$)$_4$ (6 mol%) K$_3$PO$_4$ (3 equiv.)	Dioxane/H$_2$O	30

Further switch of the base to Cs$_2$CO$_3$ (entry 2) significantly increased the amount of isolated product **158** to 46%, which follows the trend of Table 23 but with a stronger improvement. A change of the catalyst system to Pd(PPh$_3$)$_4$ and K$_3$PO$_4$ in dioxane/H$_2$O (entry 3), unexpectedly reduced the isolated yield to 30%, which is in contrast to the results obtained for the monobromide **17** in Table 23. For both the target compound **158** and the mono-coupled side product **159**, the molecular structures could be verified by single crystal analysis. **158** shows a centrosymmetric structure (Figure 71) with an angle of 39.5° between the pyridyl plane and the adjacent [2.2]paracyclophane deck plane which is less twisted than for the monopyridyl **157** (43.5°). The deck distance is 3.08 Å, which is identical to **157**.

Figure 71. Molecular structure of pseudo-*para* bis 4-pyridyl[2.2]paracyclophane (**158**) drawn at 50% probability level.

The side product **159** shows a twist angle of 40.8° in the single crystal (Figure 72) between the [2.2]paracyclophane scaffold and the pyridyl substituent, which is in the range of the bis- and monopyridyl analogues. The deck distance is slightly reduced to 3.05 Å compared to 3.08 Å of **157** and **156**, respectively.

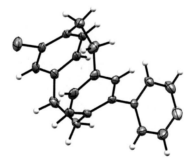

Figure 72. Molecular structure of pseudo-*para* bromo 4-pyridyl[2.2]paracyclophane (**159**) side product drawn at 50% probability level.

3.5.2   Spectroscopic Results

The racemic 4-pyridyl[2.2]paracyclophane **157** was provided to the Biedermann group to conduct all
further spectroscopic characterization methods, which were performed by Dr. Stephan Sinn.

In order to improve the binding affinity to **CB8**, **157** was methylated in dichloromethane by addition of
an excess of methyl iodide to obtain the guest **155** in 55% yield (Scheme 88).

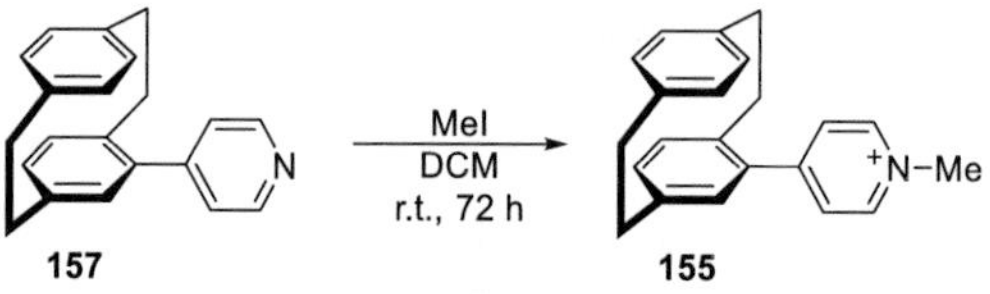

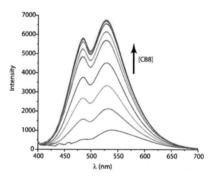

Scheme 88. Methylation to the cationic [2.2]paracyclophane guest **155**.

Indeed the binding of **155** in **CB8** can be monitored spectroscopically by excitation at 335 nm and
following the emission at 535 nm (Figure 73). After verifying that the binding mode is 1:1 and the
binding affinity amounts to log K_a = 12.59,[220] which is high enough to competitively bind with the
Alzheimer's drug memantine, an indicator displacement assay was tested.

Figure 73. Emission spectrum of **155** upon increased addition of **CB8**, λ_{exc} = 335 nm. Results and
image were provided by Dr. Sinn.[220]

Therefore, the guest **155** and the host **CB8** were prepared in blood serum, which is a challenging medium
because of competitive binders and spectroscopic noise (Figure 74). Upon titration of the Alzheimer's
drug memantine the preformed supramolecular complex CB8⊃**155** was expected to be displaced by
memantine, which shows a higher affinity of log K_a = 12.92.

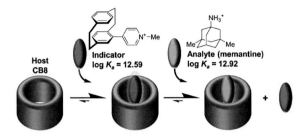

Figure 74. Indicator displacement assay showing the tested indicator/analyte/host system for the Alzheimer's drug memantine. Log K_a values were determined by Dr. Sinn.[220]

Fortunately, this trend was observed (Figure 75). The linear trend is caused by the strong binding affinity of **155**, which is strong enough to exclude all other binding interferents present in blood serum. This is observed at a µmolar memantine range of $0 - 8.3$ µM, which is within the therapeutically relevant concentration range of 0.03 to 1.00 µM in blood plasma.[203-204]

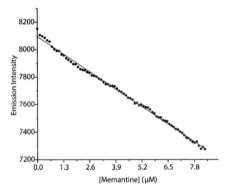

Figure 75. Linearly decreasing emission intensity of **CB8⊃155** (λ_{exc} = 335 nm, λ_{em} = 535 nm, c = 9.9 µM) plotted over increasing memantine concentrations in the µmolar range ($0 - 8.3$ µM) in blood serum at 25 °C. Results and image were provided by Dr. Sinn.[220]

3.5.3   Chiral Chromatography of Starting Materials

Motivated by these results and the drawback that the sensitivity in blood serum is hampered by spectroscopic noise of the medium, enantioenriched derivatives of **155** were needed in order to exploit the fluorescence-detected circular dichroism (FDCD), which would allow to selectively measure chiral **155** without the interference of non-fluorescent or non-chiral spectroscopic interferants present in the medium.

For this purpose, the enantioenriched 4-bromo[2.2]paracyclophanes (**17**) separated by HPLC with chiral phases (see chapter 3.2.4 and Figure 48) were subjected to the previously optimized cross-coupling to obtain 4-pyridyl[2.2]paracyclophanes **157** in an enantio-enriched manner. The enantiomeric excess of these guest precursors was determined *via* chiral analytical HPLC to 99%*ee* for (R_P)-**157** and 93%*ee* for (S_P)-**157** (Figure 76). The difference in enantiomeric excess is caused by tailing of the (R_P)-**17** eluent fraction which contaminates the second eluent fraction containing (S_P)-**17** during preparative chiral HPLC.

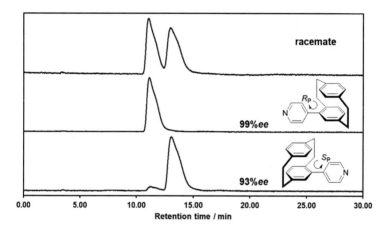

Figure 76. Analytical HPLC run of enantioenriched 4-pyridyl[2.2]paracyclophanes (**157**) on a OD-H column using 2% isopropanol in *n*-hexane with a flow rate of 1 mL/min. Detection was performed at 330 nm.

Methylation of each enantiomer was conducted according to the procedure provided by the Biedermann group (Scheme 88) to obtain both methylated enantiomers (R_P)-**155** and (S_P)-**155.**

4   Summary and Outlook

In this work, many new synthetic routes are presented giving access to novel [2.2]paracyclophane derivatives, targeting specific current questions of material sciences. New emitters showing thermally activated delayed fluorescence (TADF), both through-space and *via* increased twist caused by the considerable steric demand of the (1,4)carbazolophane subunit were designed.

A verification of chirality transfer from CVD monomers to chiral parylene fibers was enabled by the synthesis of suitable monomers and the predicted host properties of 4-(4'-pyridyl)[2.2]paracyclophane could be determined by an optimization of the respective Suzuki-Miyaura cross-coupling protocol.

Finally, new peculiarities of the [2.2]paracyclophane were observed and evaluated such as *para*-selective C–H activation and (1,7)carbazolophane formation.

4.1   Through-Space Conjugated TADF Emitters

To investigate the [2.2]paracyclophane as a through-space bridging unit for TADF, two acceptor structures, namely the benzoyl (Figure 77) and the nitrile (Figure 79) acceptor units. In the first part, the pseudo-*geminal* and pseudo-*para* bromo benzoyl intermediates were synthesized. While the synthesis of **87** relied on a transannular directive electrophilic aromatic substitution tendency of the benzoyl, the relative orientation of **93** was defined by isolation of the pseudo-*para* dibromide intermediate.

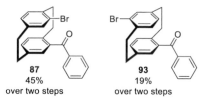

Figure 77. Overview of the synthesized bromo benzoyl acceptor unit intermediates.

From these bromo intermediates, Suzuki-Miyaura cross-coupling protocols were tested to obtain triphenylamine (**91** and **95**, Figure 78) and 4-*N*-carbazolyl-phenyl derivatives. In all cases moderate yields were obtained for the optimized Suzuki-Miyaura cross-coupling reactions. Since only the triphenylamine-based emitters were luminescent, their spectroscopic properties were determined in collaboration with the Zysman-Colman group from the University of St. Andrews. Both emitters showed nearly identical emission wavelengths of 480 nm and 465 nm for **91** and **93**, respectively. The determined delayed component of emission was in the short microsecond regime 1.8–3.6 μs, supporting the presence of TADF. As the photoluminescence quantum yields were low with 12–15%, no devices were fabricated. DFT calculations showed that the HOMO and LUMO are located on precisely one of the phenyl rings of the [2.2]paracyclophane, further corroborating the through-space electronic communication by this scaffold.

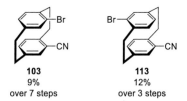

91
65%
λ_{PL} = 480 nm
Φ_{PL} = 12%
τ_D = 1.8 µs

95
65%
λ_{PL} = 465 nm
Φ_{PL} = 15%
τ_D = 3.6 µs

Figure 78. Structures of the TADF emissive through-space benzoyl acceptor-based emitters. HOMO distribution is indicated in blue, LUMO distribution is indicated in red, and the thin film spectroscopic properties are summarized.

In the second part, the benzoyl acceptor units were replaced by nitriles (Figure 79). To access the nitrile group, a new method was developed focusing on dehydration of an aldoxime intermediate obtained from an aldehyde function. While the pseudo-*para* intermediate **113** could be synthesized by the same synthetic strategy as the benzoyl derivative **93** in three steps, the synthetic access to the pseudo-*geminal* bromo nitrile **103** posed a bigger challenge. The key intermediate **103** was obtained in a seven step synthesis with formation of a suitable functional group, that first directs the electrophilic aromatic bromination to the pseudo-*geminal* position, and then can be synthetically converted to the aldehyde, which was subjected to aldoxime formation followed by dehydration in a one-pot procedure.

103
9%
over 7 steps

113
12%
over 3 steps

Figure 79. Overview of the synthesized bromo nitrile acceptor unit intermediates.

With these intermediates in hand, different Suzuki-Miyaura cross-coupling protocols were tested to obtain triphenylamino, 4-(*N*-carbazolyl)phenyl and 4-(*N*-acridan)-phenyl donor based emitters. Interestingly, the expected steric hindrance of the pseudo-*geminal* bromo nitrile **103**, when compared to the less hindered isomer **113**, was much more pronounced than for the benzoyls. Again, only the triphenylamino donor set (**107** and **114**, Figure 80) showed photoemission. Their spectroscopic properties were determined in collaboration with the Zysman-Colman group from the University of St. Andrews and showed that no distinct delayed component is present with the emitter **114** showing a delayed component in the sub-microsecond range. The emission wavelengths were determined in the deep-blue region at 426 nm and 410 nm. Since no TADF component was observed, no further investigation was conducted on this set of molecules. DFT calculation confirmed the absence of TADF due to a significant change of the LUMO location when compared to the benzoyls. Apparently, due to the small size of the nitrile, the LUMO is distributed over both [2.2]paracyclophane rings and therefore overlaps with the HOMO. This is indicated with color in Figure 80.

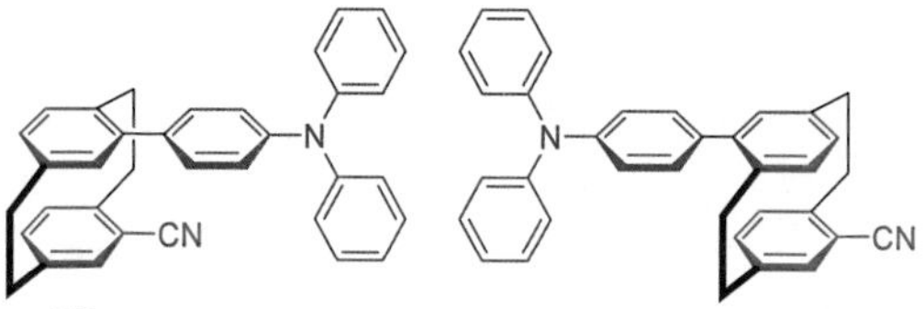

107
63%
λ_{PL} = 426 nm
Φ_{PL} = 52%

114
93%
λ_{PL} = 410 nm
Φ_{PL} = 42%

Figure 80. Structures of the non-TADF emissive nitrile acceptor-based emitters. HOMO distribution is indicated in blue, LUMO distribution is indicated in red and the thin film spectroscopic properties are summarized.

4.1.1 Outlook for Thorough-Space Emitters

As the benzoyls verified their TADF properties, no further investigation is needed, but the nitrile emitters **107** and **114** with their deep-blue emission and high photoluminescence quantum yields are suitable substrates for further derivatization. Zysman-Colman *et al.* showed that nitriles can be converted to oxadiazols, which in turn shift the emission spectrum further to the blue region and allow the LUMO to be distributed over more atoms and therefore away from the HOMO.[128]

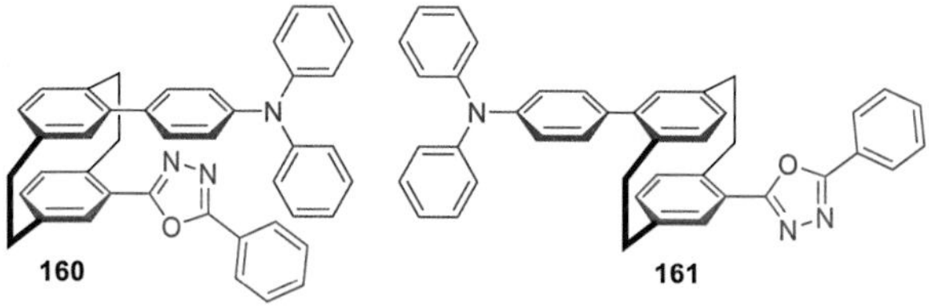

160 **161**

Figure 81. Proposed derivatization of fluorescent emitters **107** and **114** to TADF emitters **160** and **161** exhibiting an oxadiazol acceptor unit.

4.2   Carbazolophane Donor for TADF Turn-On

In the next project, the 4-(N-carbazolyl)-triphenyl triazine emitter system **76** was taken as a literature known parent compound, which is fluorescent, but does not show TADF. Additional twist along the C–N bond of the carbazole donor unit does introduce as large enough separation of HOMO and LUMO, such that ΔE_{ST} decreases enough to allow reverse intersystem crossing (RISC). A bulkier analogue of the carbazole donor unit (blue part of **76**) was needed with the additional hindrance close to the C–N bond. Common carbazole derivatives only exhibit functionalizations at the 3 position. Therefore the carbazolophane (blue part of **117**) was synthesized according to an optimized protocol based on Bolm *et al.*[39] The emitter **117** and the N-carbazolyl extended emitter **131** were obtained in a four step synthesis in 20% and 1.5%, respectively. Unfortunately, emitter **131** could not be thoroughly investigated since it was too heavy for sublimation, which was needed to ensure that purity standards are met for photophysical and device characterization by the Zysman-Colman group. Comparison of photophysical and device data from different laboratories is a challenging task, therefore the literature known emitter **76** was characterized together with the novel emitter **117**.

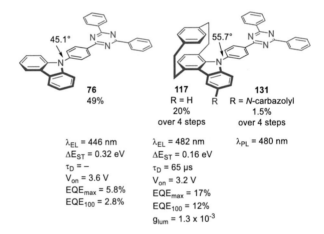

Figure 82. Structures of the synthesized **76** parent structure and the carbazolophane donor derivatives. HOMO distribution is indicated in blue, LUMO distribution is indicated in red. Twist angles were determined from single crystal analysis. The thin film spectroscopic properties and OLED performances are summarized.

While the parent compound **76** shows blue emission at 446 nm, the carbazolophane donor-based emitter **117** exhibits a bathochromic shift to the sky-blue region (482 nm). As predicted, the introduction of the carbazolophane donor with the ethyl bridge in 1-position significantly enhanced the steric demand. From crystal structure the twist was determined for both emitters to be 45.1° which was increased to 55.7° for **76** and **117**, respectively. The presence of TADF is further corroborated by the decreased ΔE_{ST} of

0.32 eV for **76** to 0.16 eV for **117** and the presence of a delayed emission lifetime of 65 µs. OLEDs were manufactured, further showing a significant increase in device performance by the presence of TADF, with an EQE$_{max}$ rising from 5.8% for the purely luminescent emitter **76** to 17% for the TADF emitter **117**. Next, the chiral emitter **117** was prepared in an enantiopure manner by separation of the carbazolophane precursors through chiral semipreparative HPLC. After determination of the absolute configuration, the respective chiral emitters (S_P)-**117** and (R_P)-**117** were prepared and showed a g$_{lum}$ value of 1.3×10^{-3}.

In summary the (1,4)carbazolophane donor unit was synthesized and successfully employed in the turn-on of the TADF relaxation mechanism enabled by the design concept of increased strain.

Among the free and the *N*-carbazolyl-extended (1,4)carbazolophanes used to synthesize emitter **117** and **131**. The chemistry in general was explored as well. Intriguingly, when trying to oxidatively cyclize **133** a novel (1,7)carbazolophane **135** was formed selectively (Scheme 89). From the crystal structure it was determined that the two annulated phenyl rings on the carbazole subunit are convoluted to 38°.

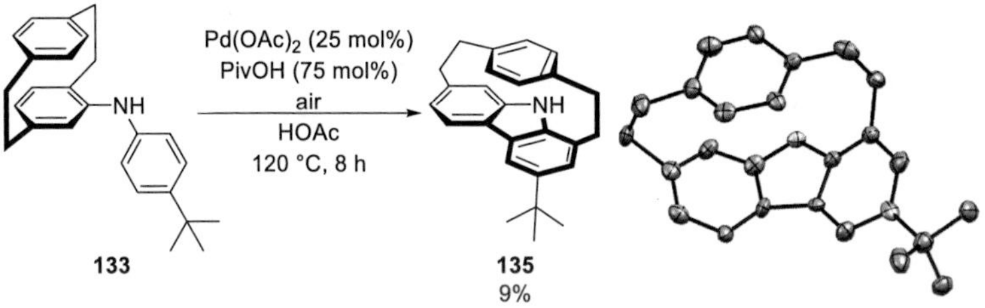

Scheme 89. Unexpected synthesis of a (1,7)carbazolophane with molecular structure.

4.2.1   Outlook for Carbazolophanes

The here employed (1,4)carbazolophane as a novel donor unit for the triazine TADF emitter system can be distributed to vary numerous other TADF systems. In particular multidonor systems pose a significant challenge, since the introduction of numerous stereogenic centers would yield diastereomers, which could influence fine-tuning of TADF properties within a given emitter class. Chirality and stereochemistry in general influence on purely organic emitter design is vastly unexplored. Additionally, the herein presented motif could further be derivatized or dimerized in order to obtain more sophisticated systems.

Additionally, the serendipitously obtained (1,7)carbazolophane **135** presents another very intriguing donor class with considerable convolution of the carbazole subunit. This donor will soon be explored and compared to the **76** emitter system.

4.3   Unexpected C–H Activation

Upon exploring the Hartwig-Buchwald cross-coupling behavior of 4-*N*-phenylamino[2.2]paracyclo-phane (**36**), a C–H activation was observed at the *para*-position which was confirmed by crystallography (Scheme 90). The reaction protocol was optimized with respect to temperature, catalyst, ligand, base, base equivalents and solvent. Despite this extensive screening of parameters, only moderate yields were obtained.

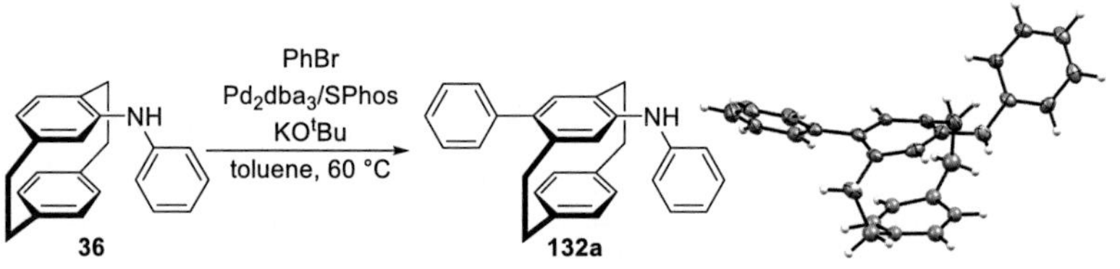

Scheme 90. Initial observation of *para* C–H activation.

The scope and limitations of the optimized reaction conditions was tested for a variety of derivatives. While no reactivity leading to the desired *para*-functionalized derivatives was observed for electron poor arenes and heteroarenes, this protocol is able to yield electron rich and sterically hindered arenes as shown in Figure 83.

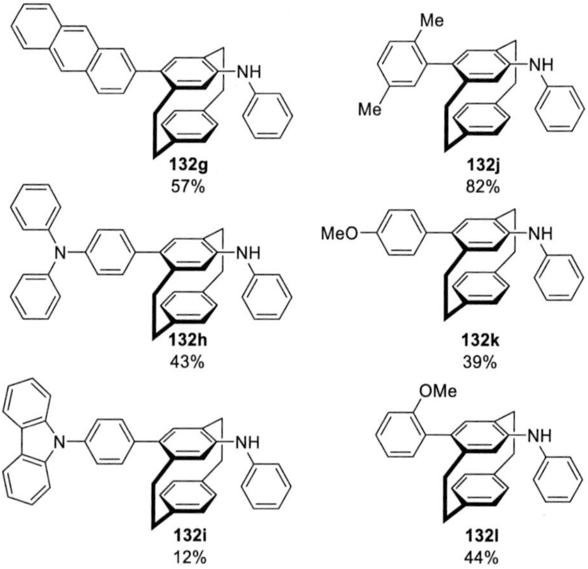

Figure 83. Overview of isolated *para*-functionalized [2.2]paracyclophanes obtained *via* optimized C–H activation protocol.

Based on thorough literature research, a mechanism for this new type of C–H activation was proposed. The electron-rich phane deck initially forms a η^6-complex with the palladium(II)-arene species, which collapses selectively to a σ-bound complex in *para*-position to the secondary amine to form a quinoimine intermediate before it undergoes reductive elimination.

In summary this newly developed protocol gives access to the less explored *para*-substituted derivatives without a *de novo* synthesis of the [2.2]paracyclophane scaffold.

4.3.1   Outlook for C–H Activation

The lack of reactivity of electron poor haloarenes in this reaction limits its use in the preparation of TADF systems. Yet, the present reactivity with electron rich haloarenes does give access to chiral aromatic compounds which are desirable structures for host materials. This is depicted in Figure 84 with CBP (**162**) as a representative. Numerous derivatives such as **163** and mixed donor systems are accessible for fine-tuning the triplet energy level, which is crucial for host materials.

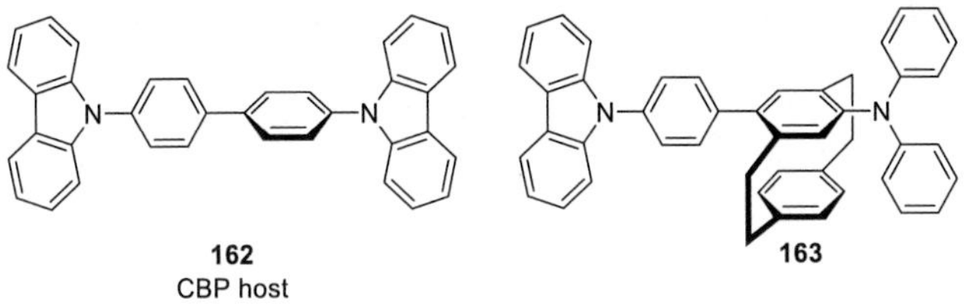

162
CBP host

163

Figure 84. Proposed host material for OLEDs with CBP shown as a representative of this compound class.

4.4   Chiral Chemical Vapor Deposition

A synthetic access to the enantiopure diastereomers **139** and **141** was established, which are suitable monomers for chemical vapor deposition (CVD) polymerization in order to tackle the question, whether liquid-crystal templated CVD would yield nanofibers with helicity induced from the prepared monomers. The initial synthetic attempt focused on stereoselective Corey-Itsuno reduction of an acetyl precursor. Since this methodology did not yield sufficient enantiomeric excesses, another route was explored focusing on initial isolation of the enantiopure (S_P)-aldehyde **5** *via* fractional crystallization of a diastereomeric imine intermediate and subsequent non-selective alkylation (Scheme 91). The absolute configuration of both isomers was unambiguously determined by crystallography and NOESY experiments.

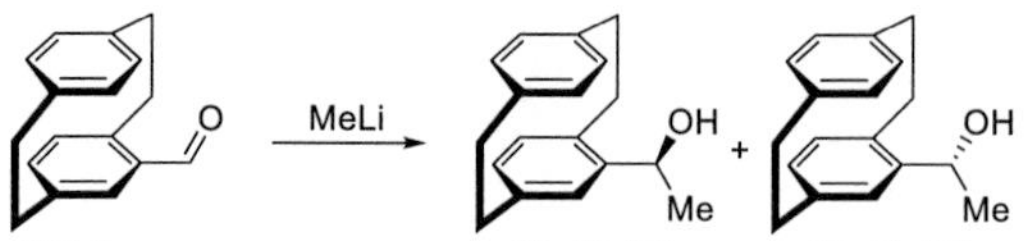

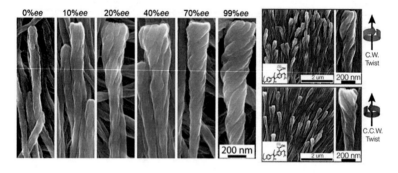

Scheme 91. Synthesized enantiopure and diastereomerically pure alcohols **139** and **141**.

CVD studies were conducted in collaboration with Prof. Lahann and Dr. Cheng at the University of Michigan. As shown in Figure 85, there is a clear influence of the stereogenic center present from the CVD presursors **139** and **141** on the sense and degree of helicity.

Figure 85. Influence enantiomeric excess and chirality on the fiber formation in liquid-crystal templated CVD, SEM images provided by Dr. Cheng.[105]

Methylated derivatives of **139** and **141** were prepared to further evaluate the role of the hydroxy functionality. Only brittle fibers were obtained, which still presented an exclusive sense of helicity depending on the isomer used. This corroborates that the mode of chiral induction proceeds *via* steric hindrance rather than by hydrogen bonding. Even though not necessary for inducing helicity, hydrogen bonding turned out to be crucial to obtain stable fibers in the liquid crystal templated CVD protocol. In summary, enantiopure and diastereomerically pure monomers were obtained and the absolute configuration was unambiguously assigned. The stereogenic center showed to be responsible for the induction of fiber helicity in non-chiral liquid crystal templated CVD, which is an example of chirality transfer from a molecule to the nanometer environment.

4.4.1   Outlook for Chiral CVD

Clearly the scope of suitable chiral monomers can be broadened in order to thoroughly evaluate whether the influence of helicity induction is caused by bulkiness, or a pre-organization *via* hydrogen bonding. Additionally, co-polymerization of helicity inducing monomers with suitable monomers for post-functionalization can be evaluated to place functionality within the well-defined helicity on the nanometer scale.

4.5   [2.2]Paracyclophane-based Molecular Sensor

Ultimately, a Suzuki-Miyaura cross-coupling protocol was developed to obtain 4-(4'-pyridyl)[2.2]para-cyclophane (**157**) in good yield (Scheme 92.). This was done to further advance the knowledge of cross-coupling behavior of the challenging 4-bromo[2.2]paracyclophane (**17**) and the pseudo-*para* dibromo derivative.

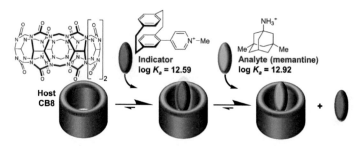

Scheme 92. Optimized Suzuki-Miyaura cross-coupling protocol for 4-pydriyl boronic acid.

The structural parameters of **157** indicated, that this compound is a suitable guest molecule for the cucurbit[8]uril (**CB8**) host in supramolecular chemistry. To corroborate this, the Biedermann group experimentally determined the binding affinity of the methylated cation of **157** experimentally to log K_a = 12.59. This strong binding affinity allowed an indicator displacement assay (IDA) to determine the concentration of the Alzheimer's drug memantine in a therapeutically relevant concentration range of 0 – 8.3 µM in blood serum. The [2.2]paracyclophane analyte was detected *via* fluorescence at 535 nm.

Scheme 93. Indicator displacement assay (IDA) performed by the Biedermann group.[220]

To further increase the sensitivity, **157** was synthesized in an enantiopure manner by chiral resolution of bromide **17** *via* chiral HPLC.

4.5.1   Outlook for [2.2]paracyclophane-based Molecular Sensors

As blood serum is a challenging medium not only with regard to competitively binding molecules, but also because of the presence of spectroscopically active molecules in the medium, a bathochromic shift of the read-out fluorescence wavelength is desirable. To tackle this, the C–H methodology presented above would be suitable to influence the emission wavelength.

5 Experimental Section

5.1 General Remarks

5.1.1 Nomenclature of [2.2]Paracyclophanes

The IUPAC nomenclature for cyclophanes in general is rather confusing. Therefore Vögtle *et al.* developed a specific cyclophane nomenclature, which is based on a core-substituent ranking.[23] This is exemplified in Figure 86 for the [2.2]paracyclophane.

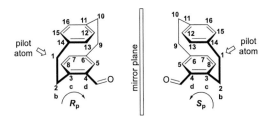

Figure 86. Nomenclature illustrated on the enantiomers of 4-formyl[2.2]paracyclophane.

The core structure is named according to the length of the aliphatic bridges in squared brackets (e.g. [n.m]) and the benzene substitution patterns (*ortho, meta* or *para*). [2.2]Paracyclophane belongs to the D_{2h} symmetry, which is broken by the first substituent, resulting in two planar chiral enantiomers. They cannot be drawn in a racemic fashion. By definition, the arene bearing the substituent is set to a chirality plane, and the first atom of the cyclophane structure outside the plane and closest to the chirality center is defined as the "*pilot atom*". If both arenes are substituted, the substituent with higher priority according to the Cahn-Ingold-Prelog (CIP) nomenclature is preferred.[24] The stereo descriptor is determined by the sense of rotation viewed from the pilot atom. To describe the positions of the substituents correctly, an unambiguous numeration is needed. The numbering of the arenes follows the sense of rotation determined by CIP. To indicate the stereochemistry of the planar chirality, a subscripted *p* is added. Unfortunately, the numbering of the second arene is inconsistent in the literature. (In this thesis, the rotating direction of numbering of the second ring is set in a way that viewed from the center of the molecule, the sense of rotation consistent.) Therefore another description based on the benzene substitution patterns is preferred for disubstituted [2.2]paracyclophanes. Substitution on the other ring is commonly named pseudo-(*ortho, meta, para* or *geminal*). With respect to the scope and aim of the respective works an even com concise nomenclature can be used, such as "trans" for pseudo-*para* (4,16) derivatives and "cis" for pseudo-*geminal* (4,13) derivatives.

In this thesis, chiral heterocycle-fused derivatives are presented. In this case the cyclophane, and the carbazole nomenclatures are exerted simultaneously (Figure 87).

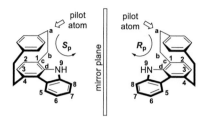

Figure 87. Nomenclature of the [2]paracyclo[2](1,4)carbazolophane (short: carbazolophane).

The first report of the [2]paracyclo[2](1,4)carbazolophane by Bolm *et al.*[39] presented the nomenclature, but no information on chirality was given The stereo descriptors follow the same standard as reported for the [2.2]paracyclophane. If viewed as a disubstituted [2.2]paracyclophane, then the nitrogen is assigned a higher priority than the carbon substituent, resulting in the respective sense of numbering as depicted in red.

5.1.2   Materials and Methods

Nuclear Magnetic Resonance Spectroscopy (NMR)

The NMR spectra of the compounds described herein were recorded on a Bruker Avance 300 NMR instrument at 300 MHz for ^{1}H NMR and 75 MHz for ^{13}C NMR, a Bruker Avance 400 NMR instrument at 400 MHz for ^{1}H NMR and 101 MHz for ^{13}C NMR, a Bruker Avance 500 NMR instrument at 500 MHz for ^{1}H NMR, 125 MHz for ^{13}C NMR and 470 MHz for ^{19}F NMR or a Bruker Avance 600 MHz instrument at 600 MHz for ^{1}H NMR and 150 MHz for ^{13}C NMR.

The NMR spectra were recorded at room temperature in deuterated solvents acquired from Eurisotop. The chemical shift δ is displayed in parts per million [ppm] and the references used were the ^{1}H and ^{13}C peaks of the solvents themselves: d_1-chloroform (CDCl$_3$): 7.26 ppm for ^{1}H and 77.0 ppm for ^{13}C, d_6-dimethyl sulfoxide (DMSO-d_6): 2.50 ppm for ^{1}H and 39.4 ppm for ^{13}C and d_6-benzene (C$_6$D$_6$): 7.16 ppm for ^{1}H and 128.06 ppm for [13]C.

For the characterization of centrosymmetric signals, the signal's median point was chosen, for multiplets the signal range. The following abbreviations were used to describe the proton splitting pattern: d = doublet, t = triplet, m = multiplet, dd = doublet of doublet, ddd = doublet of doublet of doublet, dt = doublet of triplet. Absolute values of the coupling constants "J" are given in Hertz [Hz] in absolute value and decreasing order. Signals of the ^{13}C spectrum were assigned with the help of distortionless enhancement by polarization transfer spectra DEPT90 and DEPT135 and were specified in the following way: DEPT: + = primary or tertiary carbon atoms (positive DEPT signal), – = secondary carbon atoms (negative DEPT signal), C$_{quat.}$ = quaternary carbon atoms (no DEPT signal).

Infrared Spectroscopy (IR)

The infrared spectra were recorded with a Bruker, IFS 88 instrument. Solids were measured by attenuated total reflection (ATR) method. The positions of the respective transmittance bands are given in wave numbers $\bar{v}$ [cm^{-1}] and was measured in the range from 3600 cm^{-1} to 500 cm^{-1}.

Characterization of the transmittance bands was done in sequence of transmission strength T with following abbreviations: vs (very strong, 0−9% T), s (strong, 10−39% T), m (medium, 40−69% T), w (weak, 70−89% T), vw (very weak, 90−100% T) and br (broad).

Mass Spectrometry (MS)

Electron ionization (EI) and fast atom bombardment (FAB) experiments were conducted using a Finnigan, MAT 90 (70 eV) instrument, with 3-nitrobenzyl alcohol (3-NBA) as matrix and reference for high resolution. For the interpretation of the spectra, molecular peaks [M]+, peaks of protonated molecules [M+H]$^+$ and characteristic fragment peaks are indicated with their mass-to-charge ratio (m/z) and in case of EI their intensity in percent, relative to the base peak (100%) is given. In case of high-resolution measurements, the tolerated error is 0.0005 m/z.

APCI and ESI experiments were recorded on a Q-Exactive (Orbitrap) mass spectrometer (Thermo Fisher Scientific, San Jose, CA, USA) equipped with a HESI II probe to record high resolution. The tolerated error is 5 ppm of the molecular mass. Again, the spectra were interpreted by molecular peaks [M]$^+$, peaks of protonated molecules [M+H]$^+$ and characteristic fragment peaks and indicated with their mass-to-charge ratio (m/z).

Elemental Analysis (EA)

Elemental analysis was done on a Elementar vario MICRO instrument. The weight scale used was a Sartorius M2P. Calculated (calc.) and found percentage by mass values for carbon, hydrogen, nitrogen and sulfur are indicated in fractions of 100%.

Thin Layer Chromatography (TLC)

For the analytical thin layer chromatography, TLC silica plates coated with fluorescence indicator, from Merck (silica gel 60 F254, thickness 0.2 mm) were used. UV-active compounds were detected at 254 nm and 366 nm excitation wavelength with a Heraeus UV-lamp, model Fluotest.

Weight Scale

For weightings of solids and liquids, a Sartorius model LC 620 S was used.

Solvents and Chemicals

Solvents of p.a. quality (per analysis) were commercially acquired from Sigma Aldrich, Carl Roth or Acros Fisher Scientific and, unless otherwise stated, used without further purification. Dry solvents were either purchased from Carl Roth, Acros or Sigma Aldrich (< 50 ppm H_2O over molecular sieves). All reagents were commercially acquired from abcr, Acros, Alfa Aesar, Sigma Aldrich, TCI, Chempur, Carbolution or Synchemie, or were available in the group. Unless otherwise stated, all chemicals were used without further purification.

CD Spectrometry

Circular dichroism measurements of the (S_P, R_P)-4-bromo[2.2]paracyclophane enantiomers were conducted using a Jasco J1500 CD/FDCD spectrometer in the Biedermann group. Samples were preparated in acetonitrile and an absorbance spectrum was measures beforehand using a FP8300 spectrometer from Jasco.

High Performance Liquid Chromatography (HPLC)

Purification of diastereomeric mixtures and chiral intermediates, such as the (S_P,X)-1-(4-[2.2]paracyclophanyl)ethanols, (*rac*)-carbazolophanes, and the (*rac*)-4-bromo[2.2]paracyclophane were conducted using two preparative HPLC setups.

1) JASCO HPLC System (LC-NetII/ADC) equipped with two PU-2087 Plus pumps, a CO-2060 Plus thermostat, an MD-2010 Plus diode array detector and a CHF-122SC fraction collector of ADVANTEC. For the purification of the (*rac*)-4-bromo[2.2]paracyclophane, a Daicel Chiralpak (AZ-H 20 × 250 mm, particle size of 5 μm) was used with HPLC-grade acetonitrile as mobile phase. Detection was conducted at 218 nm.

2) PuriFlash 4125 by Interchim, equipped with InterSoft V5.1.08, a UV diode array detector (200–600 nm), detection was set to 200 nm and 256 nm. In case of (*rac*)-carbazolophane, the YMC CHIRAL ART Amylose-SA (10 × 250 mm, particle size of 5 μm) was used as stationary phase. A gradient of HPLC-grade *n*-hexane/ethyl acetate; 90:10 to 80:20 was used. In case of the diastereometic mixture (S_P,X)-1-(4-[2.2]paracyclophanyl)ethanol, a simple silica cartridge was used as stationary phase, and *n*-hexane/ethyl acetate was used as mobile phase.

Analysis of the enantiomeric excess was conducted using an AGILENT HPLC 1100 series system with a G1322A degasser, a G1211A pump, a G1313A autosampler, a G1316A column oven and a G1315B diode array system. For the (*rac*, S_P, R_P)-4-bromo[2.2]paracyclophanes and (*rac*, S_P, R_P)-4-pyridyl[2.2]paracyclophanes a Chiralpak AS (4.6 × 250 mm, 5 μm particle size) and Chiralpak OD-H (4.6 × 250 mm, 5 μm particle size) columns were used with HPLC-grade *n*-hexane/isopropanol as mobile phase, respectively. For the (*rac*, S_P, R_P)-carbazolophanes and (*rac*, S_P, R_P)-carbazolophanyl-based TADF emitters a YMC CHIRAL ART Amylose-SA column (4.6 × 250 mm, 5 μm particle size) was used as stationary phase with *n*-hexane/isopropanol as mobile phase.

Gas Chromatography (GC)

To determine the degree of conversion, gas chromatograms were recorded on a Bruker 430 GC device equipped with a FactorFourTM VF-5ms (30 m × 0.25 mm × 0.25 mm) capillary column and a flame ionization detector (FID).

Experimental Procedure

Air- and moisture-sensitive reactions were carried out under argon atmosphere in previously baked out glassware using standard Schlenk techniques. Solid compounds were ground using a mortar and pestle before use, liquid reagents and solvents were injected with plastic syringes and stainless-steel cannula of different sizes, unless otherwise specified.

Reactions at low temperature were cooled using shallow vacuum flasks produced by Isotherm, Karlsruhe, filled with a water/ice mixture for 0 °C, water/ice/sodium chloride for –20 °C or isopropanol/dry ice mixture for –78 °C. For reactions at high temperature, the reaction flask was equipped with a reflux condenser and connected to the argon line.

Solvents were evaporated under reduced pressure at 40 °C using a rotary evaporator. Unless otherwise stated, solutions of inorganic salts are saturated aqueous solutions.

Reaction Monitoring

The progress of the reaction in the liquid phase was monitored by TLC. UV active compounds were detected with a UV-lamp at 254 nm and 366 nm excitation wavelength. When required, vanillin solution, potassium permanganate solution or methanolic bromocresol green solution was used as TLC-stain, followed by heating. Additionally, APCI-MS (atmospheric pressure chemical ionization mass spectrometry) was recorded on an Advion expression CMS in positive ion mode with a single quadrupole mass analyzer. The observed molecule ion is interpreted as $[M+H]^+$.

Product Purification

Unless otherwise stated, the crude compounds were purified by column chromatography. For the stationary phase of the column, silica gel, produced by Merck (silica gel 60, 0.040 × 0.063 mm, 260 – 400 mesh ASTM), and sea sand by Riedel de-Haën (baked out and washed with hydrochloric acid) were used. Solvents used were commercially acquired in HPLC-grade and individually measured volumetrically before mixing.

5.2   Analytical Data of Through-Space TADF Emitter Design

5.2.1   Benzoyl Acceptor-based Emitters

(*rac*)-4-Benzoyl[2.2]paracyclophane (86)

In a 250 mL round-bottom flask, [2.2]paracyclophane (10.4 g, 50.0 mmol, 1.00 equiv.) was dissolved in 100 mL of dichloromethane and cooled to –10 °C. A solution of benzoyl chloride (11.5 mL, 14.1 g, 100 mmol, 2.00 equiv.) and AlCl$_3$ (11.7 g, 88.0 mmol, 1.75 equiv.) in 50 mL of dichloromethane were added and stirred for 1 h. The reaction mixture was filtered through glass wool and hydrolyzed with ice. Extraction was carried out with dichloromethane (200 mL), afterwards the organic layer was washed with aqueous NaHCO$_3$ solution (200 mL) and brine (200 mL), then dried over Na$_2$SO$_4$. The solvent was removed under reduced pressure and the residue was recrystallized in ethanol to yield 12.4 g of the title compound (39.7 mmol, 79%) as colorless crystals.

R$_f$ = 0.53 (cyclohexane/ethyl acetate; 40:1). – ^{1}H NMR (400 MHz, CDCl$_3$) δ = 7.72 (d, J = 7.8 Hz, 2H), 7.56 – 7.53 (m, 1H), 7.42 (t, J = 7.7 Hz, 2H), 6.77 (d, J = 7.9 Hz, 1H), 6.71 – 6.69 (m, 2H), 6.58 – 6.55 (m, 3H), 6.35 (d, J = 7.9 Hz, 1H), 3.39 – 3.09 (m, 5H), 3.07 – 2.84 (m, 3H) ppm. – ^{13}C NMR (101 MHz, CDCl$_3$) δ = 196.68 (C$_{quat}$, CO), 141.68 (C$_{quat}$), 139.96 (C$_{quat}$) 139.37 (C$_{quat}$), 138.95 (C$_{quat}$), 136.42 (C$_{quat}$), 136.14 (+, C$_{Ar}$H), 135.77 (+, C$_{Ar}$H), 134.36 (+, C$_{Ar}$H), 132.82 (+, C$_{Ar}$H), 132.76 (+, C$_{Ar}$H), 132.52 (+, C$_{Ar}$H), 132.45 (+, C$_{Ar}$H), 131.22 (+, C$_{Ar}$H), 130.04 (+, C$_{Ar}$H), 128.31 (+, C$_{Ar}$H), 35.67 (–, CH$_2$), 35.36 (–, CH$_2$), 35.30 (–, CH$_2$), 35.17 (–, CH$_2$) ppm. – IR (ATR) $\tilde{v}$ = 2919 (w), 2848 (vw), 1649 (m), 1594 (w), 1446 (w), 1411 (vw), 1318 (w), 1294 (w), 1269 (w), 1196 (w), 976 (w), 941 (w), 907 (w), 891 (w), 835 (w), 803 (w), 726 (w), 700 (m), 656 (w), 634 (m), 509 (m), 454 (vw) cm^{-1}. – MS (EI, 70 eV) m/z (%) = 313 (25) [M+H]$^+$, 312 (100) [M]$^+$, 208 (86) [M–C$_8$H$_8$]$^+$, 207 (73) [M–C$_8$H$_9$]$^+$, 105 (11) [C$_8$H$_9$]$^+$, 104 (29) [C$_8$H$_8$]$^+$. – HRMS (EI, C$_{23}$H$_{20}$O) calc. 312.1509; found 312.1510. The analytical data is consistent with literature.[221]

(*rac*)-4-Bromo-13-benzoyl[2.2]paracyclophane (87)

A solution of bromine (2.51 mL, 7.82 g, 49.0 mmol, 1.02 equiv.) in 60 mL of dichloromethane was prepared in a dropping funnel and 5 mL of this solution were added to iron filings (50.0 mg, 960 µmol, 2 mol%) in a 500 mL three necked flask and stirred for 1 h at room temperature. Then 100 mL of dichloromethane and (*rac*)-4-benzoyl[2.2]paracyclophane (15.0 g, 48.0 mmol, 1.00 equiv.) were added and stirred for another 30 min. The remaining bromine solution was added dropwise over a period of 5 h and the mixture was stirred for 3 days. The reaction was quenched with saturated Na_2SO_3 solution (200 mL) and stirred for 30 min until full discoloration of the mixture. The organic phase was separated, washed with brine (200 mL) and dried over Na_2SO_4. The solvent was removed under reduced pressure and the residue was purified by column chromatography (silica, cyclohexane:ethyl acetate; 40:1) to yield 8.60 g of the title compound (21.9 mmol, 46%) as a white solid.

R_f = 0.23 (cyclohexane/ethyl acetate; 40:1). – ^{1}H NMR (500 MHz, CDCl$_3$) δ = 7.73 (d, J = 7.7 Hz, 2H), 7.53 (t, J = 7.4 Hz, 1H), 7.42 (t, J = 7.6 Hz, 2H), 7.28 (d, J = 1.9 Hz, 1H), 6.74 (dd, J = 7.7, 1.9 Hz, 1H), 6.68 – 6.53 (m, 4H), 3.44 (ddd, J = 12.8, 9.6, 2.3 Hz, 1H), 3.29 (ddd, J = 13.2, 9.6, 5.7 Hz, 1H), 3.18 – 3.04 (m, 3H), 3.02 – 2.94 (m, 1H), 2.88 (ddd, J = 12.9, 10.0, 2.3 Hz, 1H), 2.79 (ddd, J = 13.2, 10.0, 5.7 Hz, 1H) ppm – ^{13}C NMR (126 MHz, CDCl$_3$) δ = 195.38 (C$_{quat}$, CO), 142.33 (C$_{quat}$), 141.39 (C$_{quat}$), 140.63 (C$_{quat}$), 138.89 (C$_{quat}$, 2C), 136.94 (+, C$_{Ar}$H), 136.21 (+, C$_{Ar}$H), 135.90 (C$_{quat}$), 135.61 (+, C$_{Ar}$H), 134.95 (+, C$_{Ar}$H), 133.90 (+, C$_{Ar}$H), 131.81 (+, C$_{Ar}$H), 130.93 (+, C$_{Ar}$H), 129.78 (+, C$_{Ar}$H), 128.26 (+, C$_{Ar}$H), 127.30 (C$_{quat}$), 36.36 (–, CH$_2$), 35.12 (–, CH$_2$), 34.64 (–, CH$_2$), 33.88 (–, CH$_2$) ppm. – IR (ATR) $\tilde{v}$ = 2927 (w), 1742 (w), 1649 (m), 1587 (w), 1471 (w), 1445 (w), 1388 (w), 1268 (m), 1240 (w), 1205 (w), 1030 (w), 981 (w), 951 (w), 890 (w), 836 (w), 800 (w), 735 (m), 700 (m), 652 (w), 637 (w), 604 (w), 519 (w), 479 (w), 390 (vw) cm^{-1}. – MS (EI, 70 eV) m/z (%) = 393 (12) [M(^{81}Br)+H]$^+$, 392 (50) [M(^{81}Br)]$^+$, 391 (12) [M(^{79}Br)+H]$^+$, 390 (47) [M(^{79}Br)]$^+$, 208 (100) [M–C$_8$H$_7$Br]$^+$, 207 (91) [M–C$_8$H$_8$Br]$^+$, 106 (39) [CHOPh]$^+$. – HRMS (EI, C$_{23}$H$_{19}$^{79}BrO) calc. 390.0614; found 390.0612.

(rac)-4-Benzoyl-13-(4'-*N*-carbazolyl)phenyl[2.2]paracyclophane (89)

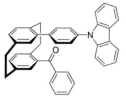

A sealable vial was charged with *(rac)*-4-bromo-13-benzoyl[2.2]paracyclophane (156.5 mg, 400 µmol, 1.00 equiv.), 4-(*N*-carbazolyl)phenyl boronic acid (229 mg, 800 µmol. 2.00 equiv.), K$_3$PO$_4$ (170 mg, 800 µmol, 2.00 equiv.), Pd(OAc)$_2$ (8.98 mg, 40.0 µmol, 10 mol%) and 2-dicyclohexylphosphino-2,6-diisopropoxybiphenyl (37.3 mg, 80.0 µmol, 20 mol%). The sealed vial was evacuated and purged with argon three times. Through the septum 7 mL of degassed toluene and 1 mL of degassed water were added, then heated to 75 °C and stirred for 16 h. The reaction mixture was diluted with 50 mL of ethyl acetate and washed with sat. aqueous NH$_4$Cl solution (3 × 30 mL) and brine (30 mL). The organic layer was dried over Na$_2$SO$_4$ and the solvent was removed under reduced pressure. The crude product was purified by column chromatography (silica, cyclohexane/ethyl acetate; 20:1) to yield 126 mg of the title compound (227 µmol, 57%) as a white solid.

R$_f$ = 0.34 (cyclohexane/ethyl acetate; 20:1). – ^{1}H NMR (400 MHz, CDCl$_3$) δ = 8.20 (d, J = 7.8 Hz, 2H), 7.64 – 7.40 (m, 11H), 7.33 (td, J = 7.4, 4.9 Hz, 4H), 7.08 (d, J = 2.0 Hz, 1H), 6.92 (dd, J = 7.8, 1.9 Hz, 1H), 6.85 (d, J = 8.1 Hz, 1H), 6.80 (d, J = 7.7 Hz, 1H), 6.66 (d, J = 2.1 Hz, 2H), 3.88 (ddd, J = 13.4, 9.3, 6.4 Hz, 1H), 3.40 (ddd, J = 12.3, 9.2, 2.5 Hz, 1H), 3.35 – 3.14 (m, 3H), 3.14 – 2.95 (m, 2H), 2.87 (ddd, J = 13.0, 9.2, 6.3 Hz, 1H) ppm. – ^{13}C NMR (101 MHz, CDCl$_3$) δ = 194.73 (C$_{quat}$, CO), 143.89 (C$_{quat}$), 141.42 (C$_{quat.}$), 141.28 (C$_{quat}$), 140.07 (C$_{quat}$), 139.80 (C$_{quat}$), 139.29 (C$_{quat}$), 138.96 (C$_{quat}$), 137.90 (C$_{quat}$), 136.58 (+, C$_{Ar}$H), 136.23 (C$_{quat}$), 136.17 (+, C$_{Ar}$H), 135.92 (+, C$_{Ar}$H), 134.95 (+, C$_{Ar}$H), 134.04 (C$_{quat}$), 132.59 (+, C$_{Ar}$H), 131.88 (+, C$_{Ar}$H), 131.42 (+, C$_{Ar}$H), 130.78 (+, C$_{Ar}$H), 130.15 (+, C$_{Ar}$H), 128.12 (+, C$_{Ar}$H), 127.02 (+, C$_{Ar}$H), 126.11 (+, C$_{Ar}$H), 123.49 (C$_{quat}$), 120.44 (+, C$_{Ar}$H), 119.99 (+, C$_{Ar}$H), 110.15 (+, C$_{Ar}$H), 37.25 (–, CH$_2$), 35.22 (–, CH$_2$), 35.15 (–, CH$_2$), 34.31 (–, CH$_2$) ppm. – Mp : 165–175 °C. – IR (ATR): $\tilde{v}$ = 2922 (vw), 1733 (vw), 1650 (vw), 1595 (vw), 1515 (vw), 1477 (vw), 1450 (w), 1334 (vw), 1315 (vw), 1269 (vw), 1230 (w), 978 (vw), 914 (vw), 837 (vw), 749 (vw), 723 (w), 700 (w), 632 (vw), 566 (vw), 525 (vw), 487 (vw), 424 (vw) cm^{-1} – MS (FAB, 3-NBA), *m/z*: 554 [M+H]$^+$, 553 [M]$^+$. – HRMS (FAB, 3-NBA, C$_{41}$H$_{31}$NO) calc. 553.2406; found 553.2405.

(rac)-4-Benzoyl-13-(4'-_N_-diphenylamino)phenyl[2.2]paracyclophane (91)

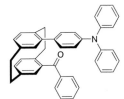

A sealable vial was charged with _(rac)_-4-bromo-13-benzoyl[2.2]paracyclophane (157 mg, 400 µmol, 1.00 equiv.), 4-(_N_-diphenylamino)phenyl boronic acid (231 mg, 800 µmol. 2.00 equiv.), K_3PO_4 (170 mg, 800 µmol, 2.00 equiv.), Pd(OAc)$_2$ (4.50 mg, 20.0 µmol, 5 mol%) and 2-dicyclohexylphosphino-2,6-diisopropoxybiphenyl (18.7 mg, 40.0 µmol, 10 mol%). The sealed vial was evacuated and purged with argon three times. Through the septum 7 mL of degassed toluene and 1 mL of degassed water were added, then heated to 75 °C and stirred for 16 h. The reaction mixture was diluted with 50 mL of ethyl acetate and washed with sat. aqueous NH$_4$Cl solution (3 × 30 mL) and brine (30 mL). The organic layer was dried over Na$_2$SO$_4$ and the solvent was removed under reduced pressure. The crude product was purified by column chromatography (silica, gradient cyclohexane/ethyl acetate; 50:1 to 20:1) to yield 145 mg of the title compound (261 µmol, 65%) as an off-white solid.

R_f =0.26 (cyclohexane/ethyl acetate; 20:1). – ^{1}H NMR (400 MHz, CDCl$_3$) δ = 7.44 (t, J = 7.3 Hz, 1H), 7.36 – 7.29 (m, 5H), 7.28 – 7.20 (m, 7H), 7.16 – 7.02 (m, 6H), 6.96 (d, J = 1.9 Hz, 1H), 6.87 (dd, J = 7.7, 1.8 Hz, 1H), 6.76 (dd, J = 11.6, 7.7 Hz, 2H), 6.57 (dd, J = 7.6, 1.9 Hz, 1H), 6.51 (d, J = 1.9 Hz, 1H), 3.82 (ddd, J = 13.3, 9.2, 6.6 Hz, 1H), 3.47 – 3.35 (m, 1H), 3.33 – 3.22 (m, 1H), 3.20 – 3.08 (m, 2H), 3.07 – 2.89 (m, 2H), 2.87 – 2.73 (m, 1H) ppm. – ^{13}C NMR (101 MHz, CDCl$_3$) δ = 194.72 (C$_{quat}$, CO), 148.10 (C$_{quat}$), 146.61 (C$_{quat}$), 144.20 (C$_{quat}$), 141.98 (C$_{quat}$), 139.98 (C$_{quat}$), 139.00 (C$_{quat}$), 138.79 (C$_{quat}$), 137.75 (C$_{quat}$), 136.46 (+, C$_{Ar}$H), 136.10 (+, C$_{Ar}$H), 135.63 (+, C$_{Ar}$H), 135.44 (C$_{quat}$), 135.20 (+, C$_{Ar}$H), 133.74 (C$_{quat}$), 131.92 (+, C$_{Ar}$H), 131.60 (+, C$_{Ar}$H), 130.81 (+, C$_{Ar}$H), 130.39 (+, C$_{Ar}$H), 130.29 (+, C$_{Ar}$H), 129.41 (+, C$_{Ar}$H), 127.92 (+, C$_{Ar}$H), 124.45 (+, C$_{Ar}$H), 124.03 (+, C$_{Ar}$H), 122.80 (+, C$_{Ar}$H), 37.32 (–, CH$_2$), 35.23 (–, CH$_2$), 35.15 (–, CH$_2$), 34.19 (–, CH$_2$) ppm. – Mp : 91–95 °C. – IR (ATR): ṽ = 2922 (vw), 1736 (w), 1649 (w), 1588 (w), 1509 (w), 1486 (w), 1445 (w), 1316 (w), 1268 (m), 1175 (w), 1074 (vw), 1046 (vw), 977 (vw), 916 (vw), 834 (w), 800 (vw), 751 (w), 696 (m), 661 (w), 648 (w), 634 (w), 550 (vw), 512 (w), 489 (vw) cm^{-1}. – MS (FAB, 3-NBA), _m/z_: 556 [M+H]$^+$, 555 [M]$^+$. – HRMS (FAB, 3-NBA, C$_{41}$H$_{33}$NO) calc. 555.2562; found 555.2561.

4,16-Dibromo[2.2]paracyclophane (22)

The synthesis followed a modified protocol reported in literature.[34] To iron powder (241 mg, 4.32 mmol, 4.50 mol%) were added 15 mL of a solution of 10.3 mL bromine (32.1 g, 201 mmol, 2.10 equiv.) in 80 mL of dichloromethane. After stirring for 1 h, the reaction mixture was diluted with 100 mL dichloromethane and [2.2]paracyclophane (20.0 g, 96.0 mmol, 1.00 equiv.) was added. The mixture was stirred for further 30 min, followed by dropwise addition of the residual bromine solution over 5 h. The reaction mixture was stirred for 3 d. Then a sat. aqueous solution of Na_2SO_3 was added to the reaction mixture, which was stirred until discoloration occurred (1 h). The organic phase was filtrated and the residual solid dried without further purification to yield 9.52 g of the title compound (26.0 mmol, 28%) as a white solid.

R_f = 0.72 (cyclohexane/ethyl acetate; 10:1). – ^{1}H NMR (400 MHz, CDCl$_3$): δ = 7.14 (dd, J = 7.8, 1.8 Hz, 2H), 6.51 (d, J = 1.7 Hz, 2H), 6.44 (d, J = 7.8 Hz, 2H), 3.49 (ddd, J = 13.0, 10.4, 2.3 Hz, 2H), 3.15 (ddd, J = 12.7, 10.4, 4.9 Hz, 2H), 2.94 (ddd, J = 12.7, 10.7, 2.3 Hz, 2H), 2.85 (ddd, J = 13.2, 10.7, 5.0 Hz, 2H) ppm. – ^{13}C NMR (101 MHz, CDCl$_3$): δ = 141.3 (C_{quat}), 138.7 (C_{quat}), 137.5 (+, C_{Ar}H), 134.2 (+, C_{Ar}H), 128.4 (+, C_{Ar}H), 126.9 (C_{quat}), 35.5 (–, CH_2), 33.0 (–, CH_2) ppm. – IR (ATR): ṽ = 2931 (vw), 2849 (vw), 1582 (vw), 1535 (vw), 1473 (vw). 1449 (vw), 1432 (vw), 1390 (w), 1185 (vw), 1030(w), 898 (w), 855 (w), 829 (w), 706 (w), 669 (w), 648 (w), 464 (w) cm^{-1}. – MS (EI, 70 eV), m/z (%): 368/366/364 (19/38/20) [M]$^+$, 184/182 (100/95) [M–C$_8$H$_7$Br]$^+$, 103 (19) [C$_8$H$_7$]$^+$. – HRMS (EI, C$_{16}$H$_{14}$^{79}Br$_2$) calc. 363.9457; found 363.9457. The analytical data is consistent with literature.[44]

(*rac*)-4 -Bromo-16-benzoyl[2.2]paracyclophane (93)

A solution of *n*-butyllithium (3.94 mL of 2.5 M in hexane, 9.84 mmol, 1.20 equiv.) was added dropwise to a solution of 4,16-dibromo[2.2]paracyclophane (3.00 g, 8.20 mmol, 1.00 equiv.) in 150 mL dry tetrahydrofuran at –78 °C under argon atmosphere. After stirring for 1 h, the reaction mixture was warmed up to 0 °C and benzoyl chloride (7.56 mL, 9.22 g, 65.6 mmol, 8.00 equiv.) was added quickly. The reaction mixture was warmed to room temperature and stirred for 16 h. The reaction mixture was quenched by addition of water and was extracted with ethyl acetate (3 × 100 mL). The combined organic phases were washed with sat. aqueous NH_4Cl solution (100 mL), brine (100 mL), dried over Na_2SO_4 and the solvent was removed under reduced pressure. The crude product was purified by column chromatography (silica, cyclohexane/ethyl acetate; 50:1) to yield 4.24 g of the title compound (5.58 mmol, 68%) as an off-white solid.

R_f = 0.22 (cyclohexane/ethyl acetate; 50:1). – ^{1}H NMR (400 MHz, CDCl$_3$) δ = 7.70 (d, J = 7.8 Hz, 2H), 7.55 (t, J = 7.4 Hz, 1H), 7.41 (t, J = 7.4 Hz, 2H), 7.35 (dd, J = 7.8, 1.9 Hz, 1H), 6.78 (dd, J = 7.9, 1.8 Hz, 1H), 6.68 (d, J = 1.9 Hz, 1H), 6.60 (d, J = 1.8 Hz, 1H), 6.56 (d, J = 7.8 Hz, 1H), 6.33 (d, J = 7.8 Hz, 1H), 3.46 (ddd, J = 13.2, 10.4, 2.7 Hz, 1H), 3.35 (ddd, J = 12.5, 10.5, 2.1 Hz, 1H), 3.28 – 3.19 (m, 2H), 3.01 – 2.93 (m, 2H), 2.92 – 2.76 (m, 2H) ppm. – ^{13}C NMR (101 MHz, CDCl$_3$) δ = 196.78 (C$_{quat}$, CO), 141.99 (C$_{quat}$), 141.19 (C$_{quat}$), 139.02 (C$_{quat}$), 138.75 (C$_{quat}$), 138.64 (C$_{quat}$), 137.09 (+, C$_{Ar}$H), 136.85 (C$_{quat}$), 134.84 (+, C$_{Ar}$H), 134.61 (+, C$_{Ar}$H), 134.48 (+, C$_{Ar}$H), 132.69 (+, C$_{Ar}$H), 132.23(+, C$_{Ar}$H), 130.16 (+, C$_{Ar}$H), 130.06 (+, C$_{Ar}$H), 128.38 (+, C$_{Ar}$H), 126.54 (C$_{quat}$), 35.11 (–, CH$_2$), 34.87 (–, CH$_2$), 34.85 (–, CH$_2$), 33.27 (–, CH$_2$) ppm. – IR (ATR): ṽ = 2926 (vw), 1645 (w), 1587 (vw), 1478 (vw), 1448 (vw), 1392 (vw), 1313 (vw), 1279 (w), 1033 (vw), 989 (vw), 908 (vw), 889 (vw), 855 (vw), 822 (w), 742 (w), 722 (vw), 701 (m), 667 (w), 652 (w), 639 (w), 514 (vw), 476 (w), 458 (vw), 397 (vw) cm^{-1} – MS (FAB, 3-NBA), *m/z*: 391/393 [M(^{79}Br/^{81}Br)+H]$^+$, 390/392 [M(^{79}Br/^{81}Br)]$^+$. – HRMS (FAB, 3-NBA, C$_{23}$H$_{19}$^{79}BrO) calc.: 390.0619; found: 390.0620. – HRMS (FAB, 3-NBA, C$_{23}$H$_{19}$^{79}BrO+H) calc. 391.0698; found 391.0700.

(rac)-4-Benzoyl-16-(4'-*N*-carbazolyl)phenyl[2.2]paracyclophane (94)

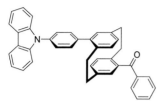

A sealable vial was charged with *(rac)*-4-bromo-16-benzoyl[2.2]paracyclophane (97.8 mg, 250 µmol, 1.00 equiv.), 4-(*N*-carbazolyl)phenyl boronic acid (144 mg, 500 µmol, 2.00 equiv.), Cs_2CO_3 (244 mg, 750 µmol, 3.00 equiv.), and $Pd(PPh_3)_2Cl_2$ (21.1 mg, 30.0 µmol, 10 mol%). The sealed vial was evacuated and purged with argon three times. Through the septum, 6 mL of degassed tetrahydrofuran and 1.5 mL of degassed water were added, then heated to 75 °C and stirred for 16 h. The reaction mixture was diluted with 50 mL of ethyl acetate and washed with sat. aqueous NH_4Cl solution (3 × 30 mL) and brine (30 mL). The organic layer was dried over Na_2SO_4 and the solvent was removed under reduced pressure. The crude product was purified by column chromatography (silica, cyclohexane/ethyl acetate; 50:1) to yield 106 mg of the title compound (191 µmol, 76%) as a white solid.

R_f = 0.14 (cyclohexane/ethyl acetate; 50:1). – ^{1}H NMR (400 MHz, CDCl$_3$) δ = 8.09 (d, J = 7.8 Hz, 2H), 7.70 – 7.64 (m, 2H), 7.60 (s, 4H), 7.49 – 7.45 (m, 3H), 7.41 – 7.31 (m, 4H), 7.25 – 7.21 (m, 2H), 6.77 – 6.69 (m, 3H), 6.67 (d, J = 1.9 Hz, 1H), 6.59 (d, J = 8.3 Hz, 1H), 6.42 (d, J = 7.8 Hz, 1H), 3.44 – 3.18 (m, 3H), 3.06 – 2.97 (m, 2H), 2.96 – 2.83 (m, 2H), 2.76 (ddd, J = 14.3, 10.2, 5.2 Hz, 1H) ppm. – ^{13}C NMR (101 MHz, CDCl$_3$) δ = 196.90 (C$_{quat}$, CO), 141.57 (C$_{quat}$), 141.29 (C$_{quat}$), 140.96 (C$_{quat}$), 140.55 (C$_{quat}$), 140.35 (C$_{quat}$), 139.60 (C$_{quat}$), 138.81 (C$_{quat}$), 136.90 (C$_{quat}$), 136.75 (C$_{quat}$), 136.60 (C$_{quat}$), 135.51 (+, C$_{Ar}$H), 134.81 (+, C$_{Ar}$H), 134.59 (+, C$_{Ar}$H), 133.22 (+, C$_{Ar}$H), 132.74 (+, C$_{Ar}$H), 131.98 (+, C$_{Ar}$H), 131.22 (+, C$_{Ar}$H), 130.97 (+, C$_{Ar}$H), 130.18 (+, C$_{Ar}$H), 128.42 (+, C$_{Ar}$H), 127.16 (+, C$_{Ar}$H), 126.11 (+, C$_{Ar}$H), 123.59 (C$_{quat}$), 120.51 (+, C$_{Ar}$H), 120.16 (+, C$_{Ar}$H), 110.04 (+, C$_{Ar}$H), 35.42 (–, CH$_2$), 35.20 (–, CH$_2$), 35.02 (–, CH$_2$), 33.28 (–, CH$_2$) ppm. – Mp : 145–150 °C. – IR (ATR): $\tilde{\upsilon}$ = 3045 (vw), 2922 (vw), 2852 (vw), 1651 (w), 1596 (vw), 1514 (w), 1478 (vw), 1449 (w), 1335 (vw), 1315 (vw), 1269 (w), 1228 (w), 1171 (vw), 1026 (vw), 913 (vw), 838 (vw), 748 (w), 723 (w), 701 (w), 656 (vw), 565 (vw), 489 (vw), 423 (vw) cm^{-1}. – MS (FAB, 3-NBA), *m/z*: 554 [M+H]$^+$, 553 [M]$^+$. – HRMS (FAB, 3-NBA, C$_{41}$H$_{31}$NO) calc. 553.2406; found 553.2403.

(rac) 4-Benzoyl-16-(4'-*N*-diphenylamino)phenyl[2.2]paracyclophane (95)

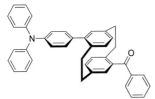

A sealable vial was charged with *(rac)*-4-bromo-16-benzoyl[2.2]paracyclophane (97.8 mg, 250 µmol, 1.00 equiv.), 4-(*N*-diphenylamino)phenyl boronic acid (144 mg, 500 µmol, 2.00 equiv.), Cs_2CO_3 (244 mg, 750 µmol, 3.00 equiv.), and $Pd(PPh_3)_2Cl_2$ (21.1 mg, 30.0 µmol, 10 mol%). The sealed vial was evacuated and purged with argon three times. Through the septum 6 mL of degassed tetrahydrofuran and 1.5 mL of degassed water were added, then heated to 75 °C and stirred for 16 h. The reaction mixture was diluted with 50 mL of ethyl acetate and washed with sat. aqueous NH_4Cl solution (3 × 30 mL) and brine (30 mL). The organic layer was dried over Na_2SO_4 and the solvent was removed under reduced pressure. The crude product was purified by column chromatography (silica, cyclohexane/ethyl acetate; 50:1) to yield 92.2 mg of the title compound (65%, 166 µmol) as an off-white solid.

R_f = 0.14 (cyclohexane/ethyl acetate; 50:1). – ^{1}H NMR (400 MHz, CDCl$_3$) δ = 7.79 – 7.69 (m, 2H), 7.59 – 7.51 (m, 1H), 7.42 (t, *J* = 7.7 Hz, 2H), 7.35 – 7.27 (m, 6H), 7.21 – 7.14 (m, 6H), 7.10 – 7.02 (m, 2H), 6.81 – 6.75 (m, 2H), 6.71 (dd, *J* = 7.8, 1.9 Hz, 1H), 6.64 (d, *J* = 1.9 Hz, 1H), 6.57 (d, *J* = 7.5 Hz, 1H), 6.43 (d, *J* = 7.8 Hz, 1H), 3.44 – 3.34 (m, 2H), 3.34 – 3.24 (m, 1H), 3.11 – 2.86 (m, 4H), 2.85 – 2.74 (m, 1H) ppm. – ^{13}C NMR (101 MHz, CDCl$_3$) δ = 196.88 (C$_{quat}$, CO), 147.84 (C$_{quat}$), 146.79 (C$_{quat}$), 141.79 (C$_{quat}$), 141.52 (C$_{quat}$), 140.00 (C$_{quat}$), 139.63 (C$_{quat}$), 138.90 (C$_{quat}$), 136.68 (C$_{quat}$), 136.64 (C$_{quat}$), 135.50 (C$_{quat}$), 135.28 (+, C$_{Ar}$H), 134.75 (+, C$_{Ar}$H), 134.53 (+, C$_{Ar}$H), 133.25 (+, C$_{Ar}$H), 132.64 (+, C$_{Ar}$H), 131.73 (+, C$_{Ar}$H), 130.56 (+, C$_{Ar}$H), 130.28 (+, C$_{Ar}$H), 130.14 (+, C$_{Ar}$H), 129.52 (+, C$_{Ar}$H), 129.45 (+, C$_{Ar}$H), 128.37 (+, C$_{Ar}$H), 124.96 (+, C$_{Ar}$H), 124.73 (+, C$_{Ar}$H), 123.41 (+, C$_{Ar}$H), 123.13 (+, C$_{Ar}$H), 35.38 (–, CH$_2$), 35.18 (–, CH$_2$), 34.89 (–, CH$_2$), 33.36 (–, CH$_2$) ppm. – Mp : 125–130 °C. – IR (ATR): ṽ = 3058 (vw), 3032 (vw), 2923 (w), 2853 (vw), 2258 (vw), 1715 (vw), 1652 (w), 1589 (w), 1487 (w), 1447 (vw), 1270 (m), 1151 (w), 1110 (w), 1024 (m), 819 (w), 753 (w), 695 (m), 617 (w), 510 (w), 483 (w), 426 (vw) cm^{-1}. – MS (FAB, 3-NBA), *m/z*: 556 [M+H]$^+$, 555 [M]$^+$. – HRMS (FAB, 3-NBA, C$_{41}$H$_{33}$NO) calc. 555.2562; found 555.2564.

5.2.2   Nitrile Acceptor-based Emitters

(*rac*)-4-Cyano[2.2]paracyclophane (97)

In a 250 mL round-bottom flask, (*rac*)-4-formyl[2.2]paracyclophane (2.36 g, 10.0 mmol, 1.00 equiv.) and hydroxylammonium chloride (2.76 g, 40.0 mmol, 4.00 equiv.) were dissolved in 100 mL of dimethylsulfoxide and stirred at 100 °C for 2 h. After cooling to room temperature, the reaction mixture was diluted with sat. NaHCO₃ solution and extracted with ethyl acetate (3 × 10 mL), then the combined organic layers were thoroughly washed with brine (3 × 20 mL) to remove DMSO. The organic layers were dried over Na₂SO₄ and the solvent was removed under reduced pressure. The crude product was purified *via* column chromatography (silica, cyclohexane/ethyl acetate; 40:1) to yield 1.84 g of the title compound (7.90 mmol, 79%) as a white solid.

R_f = 0.38 (cyclohexane/ethyl acetate 40:1). – ^{1}H NMR (400 MHz, CDCl₃) δ = 6.92 (d, *J* = 7.8 Hz, 1H), 6.77 (d, *J* = 2.0 Hz, 1H), 6.73 (dd, *J* = 7.9 Hz, *J* = 2.0 Hz, 1H), 6.59 (d, *J* = 8.0 Hz, 1H), 6.54 (s, 2H), 6.50 (d, *J* = 7.9 Hz, 1H), 3.53 (ddd, *J* = 13.4, 10.5, 3.0 Hz, 1H), 3.30 (ddd, *J* = 12.5, 10.5, 4.5 Hz, 1H), 3.21 – 3.01 (m, 6H) ppm. – ^{13}C NMR (101 MHz, CDCl₃) δ = 144.35 (C_{quat}), 141.05 (C_{quat}), 139.64 (C_{quat}), 139.23 (C_{quat}), 137.23 (+, $C_{Ar}H$), 136.89 (+, $C_{Ar}H$), 134.63 (+, $C_{Ar}H$), 133.64 (+, $C_{Ar}H$), 133.01 (+, $C_{Ar}H$), 132.82 (+, $C_{Ar}H$), 131.07 (+, $C_{Ar}H$), 119.05 (C_{quat}, $C_{Ar}CN$), 115.04 (C_{quat}, CN), 35.45 (–, CH_2), 35.20 (–, CH_2), 34.59 (–, CH_2), 34.35 (–, CH_2) ppm. – IR (ATR) $\tilde{v}$ = 2924 (w), 2851 (w), 2219 (w), 1591 (w), 1499 (w), 1484 (w), 1435 (w), 1410 (w), 1224 (vw), 945 (vw), 903 (w), 863 (w), 795 (w), 715 (m), 645 (w), 592 (m), 519 (w), 498 (m), 400 (vw) cm⁻¹. – MS (EI, 70 eV) *m/z* (%) = 234 (5) [M+H]⁺, 233 (26) [M]⁺, 104 (100) [C₈H₈]⁺·. – HRMS (EI, C₁₇H₁₅N) calc. 233.1199; found 233.1198. – EA (C₁₇H₁₅N) calc. C: 87.53, H: 6.48, N: 6.00; found C: 87.01, H: 6.64, N: 5.70. The analytical data is consistent with literature.[222]

(*rac*)-4-Carboxy[2.2]paracyclophane (105)

In a 500 mL Schlenk flask, (*rac*)-4-bromo[2.2]paracyclophane (10.g g, 35.0 mmol, 1.00 equiv.) was dissolved in 250 mL of dry tetrahydrofuran under argon atmosphere and cooled to –78 °C. Then *n*-butyllithium (14.0 mL of a 2.5 M in hexane, 35.0 mmol, 1.01 equiv.) was added slowly and the orange mixture was stirred at –78 °C for 5 h. Carbon dioxide was induced into the mixture *via* a gas frit for 1 h and the reaction mixture was stirred to room temperature for 8 h. Afterwards the reaction mixture was extracted with 1 M aqueous NaOH solution (2 × 100 mL) and washed with 100 mL dichloromethane. The aqueous phase was acidified with concentrated HCl until the product precipitated as a white solid. It was dissolved and then extracted with 200 mL of dichloromethane, washed with diluted HCl and dried over Na_2SO_4. The solvent was evaporated under reduced pressure to yield 5.69 g of the title compound (65%, 22.6 mmol) as a white solid.

R_f = 0.40 (cyclohexane/ethyl acetate 2:1). – ^{1}H NMR (400 MHz, DMSO-d_6) δ = 12.52 (s, 1H, CO$_2$H), 7.06 (d, J = 1.9 Hz, 1H), 6.70 (dd, J = 7.7, 1.9 Hz, 1H), 6.56 (d, J = 7.9 Hz, 2H), 6.51 (d, J = 7.6 Hz, 1H), 6.41 (d, J = 1.7 Hz, 2H), 4.00 (dd, J = 12.2, 9.7 Hz, 1H), 3.17 – 3.04 (m, 3H), 3.04 – 2.90 (m, 3H), 2.88 – 2.76 (m, 1H) ppm. – ^{13}C NMR (101 MHz, DMSO-d_6) δ = 167.93 (C$_{quat}$, COOH), 141.95 (C$_{quat}$), 139.53 (C$_{quat}$), 139.31 (C$_{quat}$), 139.18 (C$_{quat}$), 136.10 (+, C$_{Ar}$H), 135.90 (+, C$_{Ar}$H), 135.17 (+, C$_{Ar}$H), 133.10 (+, C$_{Ar}$H), 132.64 (+, C$_{Ar}$H), 131.97 (+, C$_{Ar}$H), 131.21 (C$_{quat}$), 131.03 (+, C$_{Ar}$H), 35.39 (–, CH$_2$), 34.61 (–, CH$_2$), 34.41 (–, CH$_2$), 26.32 (–, CH$_2$) ppm. – IR (ATR) $\tilde{v}$ = 2922 (w), 1681 (m), 1592 (w), 1555 (w), 1497 (vw), 1418 (w), 1303 (m), 1274 (m), 1200 (w), 1076 (w), 946 (w), 912 (w), 884 (w), 867 (w), 798 (w), 763 (w), 707 (w), 667 (w), 634 (w), 578 (w), 560 (vw), 511 (w), 443 (w), 399 (vw) cm^{-1}. – MS (EI, 70 eV) *m/z* (%) = 253 (7) [M]$^+$, 252 (38) [M]$^+$, 148 (19) [C$_9$H$_8$O$_2$]$^+$, 105 (100) [C$_8$H$_9$]$^+$, 104 (92) [C$_8$H$_8$]$^+$. – HRMS (EI, C$_{17}$H$_{16}$O$_2$) calc. 252.1145; found 252.1146. The analytical data is consistent with literature.[223]

(*rac*)-4-Carbomethoxy[2.2]paracyclophane (100)

In a 250 mL round-bottomed flask, (*rac*)-4-carboxy[2.2]paracyclophane (2.47 g, 9.80 mmol, 1.00 equiv.) was suspended in 150 mL of methanol. Then 6 mL of concentrated H_2SO_4 was added and the mixture was refluxed for 16 h. Upon completion, the solvent was removed under reduced pressure and the residue was dissolved in ethyl acetate (150 mL). The organic layer was washed with 1 M NaOH solution (3 × 100 mL), then with brine (100 mL). The organic layer was dried over Na_2SO_4, the solvent was removed under reduced pressure and the crude compound was purified by column chromatography (silica, cyclohexane/ethyl acetate; 10:1). To yield 2.40 g of the title compound (9.01 mmol, 92%) as a white solid.

R_f = 0.58 (cyclohexane/ethyl acetate; 10:1). – ^{1}H NMR (400 MHz, CDCl$_3$) δ = 7.14 (d, J = 2.0 Hz, 1H), 6.66 (dd, J = 7.8, 2.0 Hz, 1H), 6.58 – 6.53 (m, 2H), 6.52 – 6.43 (m, 3H), 4.09 (ddd, J = 13.0, 9.7, 1.8 Hz, 1H), 3.92 (s, 3H), 3.24 – 2.96 (m, 6H), 2.86 (ddd, J = 12.9, 10.1, 7.0 Hz, 1H) ppm. – ^{13}C NMR (101 MHz, CDCl$_3$) δ = 167.63 (C$_{quat}$), 142.74 (C$_{quat}$), 140.10 (C$_{quat}$), 139.98 (C$_{quat}$), 139.49 (C$_{quat}$), 136.55 (+, C$_{Ar}$H), 136.26 (+, C$_{Ar}$H), 135.42 (+, C$_{Ar}$H), 133.23 (+, C$_{Ar}$H), 132.84 (+, C$_{Ar}$H), 132.39 (+, C$_{Ar}$H), 131.70 (+, C$_{Ar}$H), 130.78 (C$_{quat}$), 51.92 (+, CH$_3$), 36.26 (–, CH$_2$), 35.39 (–, CH$_2$), 35.23 (–, CH$_2$), 35.09 (–, CH$_2$) ppm. – IR (ATR) ṽ = 3935 (vw), 3805 (vw), 3730 (vw), 3399 (vw), 3255 (vw), 2932 (w), 2851 (vw), 2417 (vw), 2319 (vw), 2211 (vw), 2107 (vw), 1897 (vw), 1707 (m), 1556 (w), 1432 (w), 1268 (m), 1192 (m), 1075 (m), 863 (w), 801 (w), 708 (w), 634 (w), 515 (w), 419 (vw) cm^{-1}. – HRMS (ESI, $C_{18}H_{18}O_2$) calc. 267.1380 [M+H]$^+$, found 267.1368 [M+H]$^+$; calc. 235.1117 [M–OMe]$^+$, found 235.1108 [M–OMe]$^+$. The analytical data is consistent with literature.[140]

(*rac*)-4-Bromo-13-carbomethoxy[2.2]paracyclophane (101)

A 1 L three-necked flask charged with Fe powder (554 mg, 9.92 mmol, 1 mol%) was equipped with a dropping funnel charged with bromine (2.59 mL, 8.08 g, 50.59 mmol, 1.02 equiv.) in 80 mL of dichloromethane. 5 mL of this solution was added to the iron fillings and stirred for 10 min. Subsequently, (*rac*)-4-carbomethoxy[2.2]paracyclophane (13.2 g, 49.6 mmol, 1.00 equiv.) was added as a solution in 500 mL of dichloromethane to the reaction mixture and stirred for 20 min. The remaining bromine solution was then added dropwise over 6 h to the reaction mixture and stirred for 2 days at room temperature. Saturated Na_2SO_3 solution (500 mL) was added and stirred until discoloration of the organic layer. It was extracted with dichloromethane (3 × 300 mL), the combined organic layers were washed with brine (400 mL) and the solvent was removed under reduced pressure. The crude product was purified by column chromatography (silica, cyclohexane/ethyl acetate; 20:1) to yield 8.41 g of the title compound (24.4 mmol, 49%) as a yellow solid.

R_f = 0.41 (cyclohexane/ethyl acetate; 10:1). – ^{1}H NMR (400 MHz, CDCl$_3$) δ = 7.36 (d, J = 2.0 Hz, 1H), 6.69 (dd, J = 7.7, 2.0 Hz, 1H), 6.61 (s, 1H), 6.57 – 6.53 (m, 3H), 4.48 – 4.33 (m, 1H), 3.89 (s, 3H), 3.64 – 3.48 (m, 1H), 3.18 – 2.87 (m, 6H) ppm. – ^{13}C NMR (101 MHz, CDCl$_3$) δ = 167.39 (C$_{quat}$, CO), 142.70 (C$_{quat}$), 141.31 (C$_{quat}$), 139.11 (C$_{quat}$, 2C), 136.54 (+, C$_{Ar}$H), 136.46 (+, C$_{Ar}$H), 136.08 (+, C$_{Ar}$H), 134.83 (+, C$_{Ar}$H), 134.41 (+, C$_{Ar}$H), 131.63 (+, C$_{Ar}$H), 128.53 (C$_{quat}$), 127.22 (C$_{quat}$), 51.71 (+, CH$_3$), 35.25 (–, CH$_2$), 34.98 (–, CH$_2$), 34.61 (–, CH$_2$), 33.62 (–, CH$_2$) ppm. – IR (ATR) $\tilde{v}$ = 2928 (w), 1705 (m), 1584 (w), 1432 (w), 1391 (w), 1321 (w), 1291 (w), 1264 (m), 1193 (m), 1115 (w), 1074 (m), 1031 (w), 990 (w), 956 (w), 911 (w), 882 (w), 840 (w), 776 (w), 707 (w), 667 (w), 644 (w), 593 (w), 521 (w), 476 (w) cm^{-1}. – HRMS (ESI, C$_{18}$H$_{17}$BrO$_2$) calc. 345.0485 [M(^{79}Br)+H]$^+$, found 345.0477 [M(^{79}Br)]$^+$; calc. 347.0464 [M(^{81}Br)+H]$^+$, found 347.0456 [M(^{81}Br)+H]$^+$. – EA (C$_{18}$H$_{17}$BrO$_2$) calc. C:62.62, H: 4.96; C:63,13, H: 5.05. The analytical data is consistent with literature.[140]

(*rac*)-4-Bromo-13-hydroxymethyl[2.2]paracyclophane (106)

 To a 250 mL Schlenk flask equipped with a septum, (*rac*)-4-bromo-13-carbomethoxy[2.2]paracyclophane (1.95 g, 5.66 mmol, 1.00 equiv.) was added, evacuated, and purged with argon three times. Then 100 mL of tetrahydrofurane was added and the solution was cooled to 0 °C. Subsequently, lithium aluminium hydride (472 mg, 12.5 mmol, 2.20 equiv.) was added portion-wise and the reaction mixture was stirred for 2 h. The completion of the reaction was confirmed by thin layer chromatography. The mixture was quenched with methanol, then water. The reaction mixture was diluted with saturated NH_4Cl solution and extracted with ethyl acetate (3 × 150 mL). The combined organic layers were dried over Na_2SO_4 and the solvent was removed under reduced pressure. The crude product was purified by short column chromatography (silica, cyclohexane/ethyl acetate; 5:1) to yield 1.70 g of the title compound (5.34 mmol, 95%) as a white solid.

R_f = 0.38 (cyclohexane/ethyl acetate; 5:1). – ^{1}H NMR (500 MHz, DMSO-d_6) δ = 6.65 (s, 1H), 6.62 (d, J = 7.7 Hz, 1H), 6.57 (d, J = 7.8 Hz, 1H), 6.51 (s, 2H), 6.48 (s, 1H), 4.87 (dd, J = 12.7, 6.0 Hz, 1H), 4.67 (t, J = 5.6 Hz, 1H, OH), 4.47 (dd, J = 12.7, 5.2 Hz, 1H), 3.59-3.28 (m, 2H), 3.15-2.88 (m, 6H) ppm. – ^{13}C NMR (126 MHz, DMSO-d_6) δ = 142.22 (C_{quat}), 141.60 (C_{quat}), 138.70 (C_{quat}), 137.92 (C_{quat}), 136.69 (C_{quat}), 135.25 (+, $C_{Ar}H$), 135.23 (+, $C_{Ar}H$), 134.51 (+, $C_{Ar}H$), 132.76 (+, $C_{Ar}H$), 131.88 (+, $C_{Ar}H$), 131.78(+, $C_{Ar}H$), 124.19 (C_{quat}), 62.52 (–, CH_2OH), 34.41 (–, CH_2), 34.04 (–, CH_2), 33.97 (–, CH_2), 31.19 (–, CH_2) ppm. – IR (ATR) ṽ = 3379 (vw), 2924 (w), 2850 (w), 1586 (w), 1538 (vw), 1474 (w), 1448 (w), 1390 (w), 1189 (w), 1093 (vw), 1029 (m), 996 (w), 951 (w), 907 (w), 873 (w), 833 (w), 813 (w), 761 (w), 705 (w), 668 (w), 651 (w), 623 (w), 577 (w), 515 (w), 461 (w), 386 (vw) cm^{-1}. – MS (EI, 70 eV) *m/z* (%) = 318 (22) $[M(^{81}Br)]^+$, 316 (22) $[M(^{79}Br)]^+$, 238 (32) $[C_{17}H_{18}O]^+$, 105 (78) $[C_8H_9]^+$, 104 (39) $[C_8H_8]^+$. – HRMS (EI, $C_{17}H_{17}O^{79}Br$): calc. 316.0463, found 316.0464. The analytical data is consistent with literature.[140]

(rac)-4-Bromo-13-formyl[2.2]paracyclophane (102)

A 250 mL round-bottom flask was charged with oxalyl chloride (1.18 mL, 1.75 g, 13.8 mmol, 1.10 equiv.) and 50 mL of dry dichloromethane under argon atmosphere. The solution was cooled to –78 °C and dimethylsulfoxide (2.13 mL, 2.34 g, 30.0 mmol, 2.40 equiv.) was added *via* syringe and the resulting mixture was stirred for 15 min. Subsequently, *(rac)*-4-bromo-13-hydroxymethyl[2.2]paracyclophane (3.97 g, 12.5 mmol, 1.00 equiv.) was added dropwise as a solution in 50 mL of dry dichloromethane. The resulting mixture was stirred for 15 min and triethylamine (8.66 mL, 6.32 g, 62.5 mmol, 5.00 equiv.) was added. The resulting reaction mixture was stirred for 1 h at –78 °C before allowing the reaction mixture to warm to room temperature. It was quenched with water (50 mL) and extracted with dichloromethane (3 × 100 mL). The combined organic layers were washed with brine (3 × 150 mL), dried over Na_2SO_4, and the solvent was removed under reduced pressure. The crude compound was purified by column chromatography (silica, cyclohexane/ethyl acetate; 20:1) to yield 2.49 g of the title compound (7.90 mmol, 63%) as a white solid.

R_f = 0.40 (cyclohexane/ethyl acetate; 20:1). – ^{1}H NMR (400 MHz, CDCl₃) δ = 10.30 (s, 1H, CHO), 7.11 (d, J = 2.0 Hz, 1H), 6.77 (dd, J = 7.8, 1.9 Hz, 1H), 6.66 – 6.47 (m, 4H), 4.15 (ddd, J = 13.0, 10.3, 4.2 Hz, 1H), 3.55 (ddd, J = 13.5, 10.3, 3.1 Hz, 1H), 3.22 – 2.89 (m, 6H) ppm. – ^{13}C NMR (101 MHz, CDCl₃) δ = 190.69 (+, CHO), 142.81 (C_{quat}), 141.54 (C_{quat}), 139.95 (C_{quat}), 138.69 (C_{quat}), 137.94 (+, $C_{Ar}H$), 136.68 (+, $C_{Ar}H$), 136.23 (+, $C_{Ar}H$), 135.88 (C_{quat}), 134.99 (+, $C_{Ar}H$), 133.77 (+, $C_{Ar}H$), 131.31 (+, $C_{Ar}H$), 128.42 (C_{quat}), 35.43 (–, CH_2), 34.91 (–, CH_2), 34.65 (–, CH_2), 30.21 (–, CH_2) ppm. – IR (ATR) $\tilde{v}$ = 2923 (w), 2851 (w), 1676 (m), 1583 (w), 1538 (w), 1473 (w), 1389 (w), 1280 (w), 1180 (w), 1142 (w), 1029 (w), 970 (w), 908 (w), 880 (w), 837 (w), 771 (w), 706 (w), 644 (w), 620 (w), 514 (w), 475 (vw), 456 (w), 400 (vw), 384 (vw) cm⁻¹. – MS (EI, 70 eV) *m/z* (%) = 316 (23) [M(^{81}Br)]⁺, 314 (24) [M(^{79}Br)]⁺, 184 (26) [C₈H₇(^{81}Br)]⁺, 182 (31) [C₈H₇(^{79}Br)]⁺, 132 (27) [C₉H₈O]⁺, 103 (18) [C₈H₇]⁺. – HRMS (EI, C₁₇H₁₅O^{79}Br): calc. 314.0306, found 314.0307. The analytical data is consistent with literature.[140]

(*rac*)-4-Bromo-13-cyano[2.2]paracyclophane (103)

 In a 100 mL round-bottom flask, (*rac*)-4-bromo-13-formyl[2.2]paracyclophane (2.36 g, 7.50 mmol, 1.00 equiv.) and hydroxylammonium chloride (2.61 g, 37.5 mmol, 5.00 equiv.) were dissolved in 50 mL of dimethylsulfoxide and stirred at 100 °C for 2 h. After cooling to room temperature, the reaction mixture was diluted with sat. NaHCO₃ solution and was extracted with ethyl acetate (3 × 50 mL). The combined organic layers were thoroughly washed with brine (3 × 100 mL) to remove DMSO, dried over Na₂SO₄ and the solvent was removed under reduced pressure. The crude product was purified *via* column chromatography (silica, cyclohexane/ethyl acetate; 20:1) to yield 1.26 g of the title compound (4.03 mmol, 54%) as a white solid.

R_f = 0.27 (cyclohexane/ethyl acetate; 20:1). – ¹H NMR (400 MHz, CDCl₃) δ = 6.93 (d, *J* = 1.9 Hz, 1H), 6.73 (dd, *J* = 7.9, 1.9 Hz, 1H), 6.63 (d, *J* = 1.2 Hz, 1H), 6.61 – 6.56 (m, 3H), 3.80 – 3.67 (m, 2H), 3.25 – 2.95 (m, 6H) ppm. – ¹³C NMR (101 MHz, CDCl₃) δ = 143.27 (C$_{quat}$), 141.39 (C$_{quat}$), 140.10 (C$_{quat}$), 138.52 (C$_{quat}$), 137.50 (+, C$_{Ar}$H), 136.61 (+, C$_{Ar}$H), 135.20 (+, C$_{Ar}$H), 135.18 (+, C$_{Ar}$H), 135.07 (+, C$_{Ar}$H), 131.38 (+, C$_{Ar}$H), 128.11 (C$_{quat}$), 119.16 (C$_{quat}$), 111.57 (C$_{quat}$, CN), 36.00 (–, CH₂), 34.92 (–, CH₂), 34.56 (–, CH₂), 32.19 (–, CH₂) ppm. – IR (ATR) $\tilde{v}$ = 2926 (w), 2854 (vw), 2220 (w), 1680 (vw), 1582 (w), 1538 (vw), 1472 (w), 1434 (vw), 1388 (w), 1188 (vw), 1028 (vw), 969 (vw), 929 (vw), 905 (w), 876 (w), 834 (w), 722 (w), 707 (w), 670 (w), 649 (w), 611 (w), 582 (vw), 504 (w), 485 (vw), 469 (w), 396 (vw) cm⁻¹. – MS (EI, 70 eV) *m/z* (%) = 313 (24) [M(⁸¹Br)]⁺, 311 (24) [M(⁷⁹Br)]⁺, 198 (15) [C₉H₉(⁸¹Br)]⁺, 196 (16) [C₉H₉(⁷⁹Br)]⁺, 184 (100) C₈H₇(⁸¹Br)]⁺, 182 (99) C₈H₇(⁷⁹Br)]⁺, 131 (19) [C₉H₉N]⁺, 103 (28) [C₈H₇]⁺. – HRMS (EI, C₁₇H₁₄N⁷⁹Br): calc. 311.0310, found 311.0308.

(*rac*)-4-Cyano-13-(4'-*N*,*N*-diphenylamino)phenyl[2.2]paracyclophane (107)

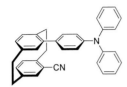 A sealable vial was charged with (*rac*)-4-bromo-13-cyano[2.2]paracyclophane (125 mg, 400 μmol, 1.00 equiv.), 4-(*N,N*-diphenylamino)phenyl boronic acid (347 mg, 1.2 mmol. 3.00 equiv.), K₃PO₄ (213 mg, 1.00 mmol, 2.50 equiv.) and Pd(PPh₃)₄ (139 mg, 120 μmol, 30 mol%). The sealed vial was evacuated and purged with argon three times. Through the septum 7 mL of degassed toluene were added, then heated to 85 °C and stirred for 60 h. The reaction mixture was diluted with 50 mL of ethyl acetate and washed with sat. aqueous NH₄Cl solution (3 × 30 mL) and brine (30 mL). The organic layer was dried over Na₂SO₄ and the solvent was removed under reduced pressure. The crude product was purified by column chromatography (silica, *n*-hexane/ethyl acetate; 20:1) to yield 106 mg of the title compound (223 μmol, 56%) as a white solid.

R_f =0.21 (*n*-hexane/ethyl acetate; 20:1). – ^{1}H NMR (400 MHz, CDCl₃) δ = 7.39 (d, J = 8.6 Hz, 2H), 7.32 – 7.25 (m, 2H), 7.26 (d, J = 3.0 Hz, 2H), 7.22 – 7.12 (m, 6H), 7.03 (t, J = 7.3 Hz, 2H), 6.89 – 6.80 (m, 2H), 6.76 – 6.66 (m, 3H), 6.52 (dd, J = 7.7, 1.9 Hz, 1H), 3.79 (ddd, J = 13.6, 9.8, 5.3 Hz, 1H), 3.36 (ddd, J = 13.3, 9.7, 3.3 Hz, 1H), 3.31 – 2.91 (m, 6H) ppm. – ^{13}C NMR (101 MHz, CDCl₃) δ = 147.85 (C_{quat}), 147.20 (C_{quat}), 144.33 (C_{quat}), 142.49 (C_{quat}), 140.59 (C_{quat}), 139.39 (C_{quat}), 136.75 (+, $C_{Ar}H$), 136.54 (C_{quat}), 136.13 (+, $C_{Ar}H$), 135.06 (+, $C_{Ar}H$), 135.02 (+, $C_{Ar}H$), 133.82 (C_{quat}), 132.05 (+, $C_{Ar}H$), 130.98 (+, $C_{Ar}H$), 130.68 (+, $C_{Ar}H$), 129.35 (+, $C_{Ar}H$), 124.73 (+, $C_{Ar}H$), 123.38 (+, $C_{Ar}H$), 122.96 (+, $C_{Ar}H$), 118.31 (C_{quat}, CN), 112.16 (C_{quat}), 35.20 (–, CH_2), 35.18 (–, CH_2), 34.49 (–, CH_2), 33.43 (–, CH_2) ppm. – IR (ATR) ṽ = 3036 (vw), 2923 (w), 2852 (vw), 2617 (vw), 2454 (vw), 2323 (vw), 2220 (vw), 1901 (vw), 1779 (vw), 1585 (w), 1485 (m), 1268 (w), 1180 (w), 1076 (vw), 836 (w), 756 (m), 695 (m), 622 (w), 505 (w) cm⁻¹. – HRMS (ESI, C₃₅H₂₈N₂) calc. 477.2325 [M+H]⁺, found 477.2300 [M+H]⁺.

(*rac*)-4-Cyano-13-(4'-*N*-carbazolyl)phenyl[2.2]paracyclophane (108)

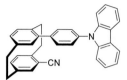

A sealable vial was charged with (*rac*)-4-bromo-13-cyano[2.2]paracyclophane (125 mg, 400 μmol, 1.00 equiv.), 4-(*N*-carbazolyl)phenyl boronic acid (230 mg, 800 μmol. 2.00 equiv.), K$_3$PO$_4$ (213 mg, 1.00 mmol, 2.50 equiv.) and Pd(PPh$_3$)$_4$ (46.2 mg, 40.0 μmol, 10 mol%). The sealed vial was evacuated and purged with argon three times. Through the septum 7 mL of degassed toluene were added, then heated to 85 °C and stirred for 60 h. The reaction mixture was diluted in 50 mL of ethyl acetate and washed with sat. aqueous NH$_4$Cl solution (3 × 30 mL) and brine (30 mL). The organic layer was dried over Na$_2$SO$_4$ and the solvent was removed under reduced pressure. The crude product was purified by column chromatography (silica, *n*-hexane/ethyl acetate; 20:1) to yield 50.0 mg of the title compound (105 mmol, 26%) as a white solid.

R$_f$ =0.24 (*n*-hexane/ethyl acetate; 20:1). – ^{1}H NMR (400 MHz, CDCl$_3$) δ = 8.16 (d, *J* = 7.8 Hz, 2H), 7.78 (d, *J* = 8.4 Hz, 2H), 7.66 (d, *J* = 8.4 Hz, 2H), 7.55 (d, *J* = 8.2 Hz, 2H), 7.44 (ddd, *J* = 8.3, 7.1, 1.3 Hz, 2H), 7.30 (td, *J* = 7.5, 1.0 Hz, 2H), 6.97 (d, *J* = 1.9 Hz, 1H), 6.89 (dd, *J* = 7.9, 1.8 Hz, 1H), 6.83 (d, *J* = 1.9 Hz, 1H), 6.79 – 6.75 (m, 2H), 6.64 – 6.57 (m, 1H), 3.82 (ddd, *J* = 13.7, 9.8, 5.3 Hz, 1H), 3.40 (ddd, *J* = 13.4, 9.8, 3.3 Hz, 1H), 3.32 – 3.20 (m, 3H), 3.19 – 2.96 (m, 3H) ppm. – ^{13}C NMR (101 MHz, CDCl$_3$) δ = 144.28 (C$_{quat}$), 142.03 (C$_{quat}$), 141.00 (C$_{quat}$), 140.77 (C$_{quat}$), 139.72 (C$_{quat}$), 138.81 (C$_{quat}$), 136.94 (+, C$_{Ar}$H), 136.85 (C$_{quat}$), 136.63 (C$_{quat}$), 136.40 (+, C$_{Ar}$H), 135.19 (+, C$_{Ar}$H), 135.15 (+, C$_{Ar}$H), 135.07 (+, C$_{Ar}$H), 132.85 (+, C$_{Ar}$H), 131.72 (+, C$_{Ar}$H), 131.03 (+, C$_{Ar}$H), 127.01 (+, C$_{Ar}$H), 126.15 (+, C$_{Ar}$H), 123.52 (C$_{quat}$), 120.32 (+, C$_{Ar}$H), 120.01 (+, C$_{Ar}$H), 118.42 (C$_{quat}$, CN), 112.32 (C$_{quat}$), 110.23 (+, C$_{Ar}$H), 35.21 (–, CH$_2$), 34.56 (–, CH$_2$), 33.33 (–, CH$_2$), 29.85 (–, CH$_2$) ppm. – IR (ATR) ṽ = 3041 (vw), 2923 (w), 2852 (w), 2678 (vw), 2395 (vw), 2320 (vw), 2217 (vw), 2044 (vw), 1897 (vw), 1721 (vw), 1594 (vw), 1515 (w), 1450 (w), 1317 (w), 1229 (w), 1016 (vw), 911 (vw), 838 (w), 748 (w), 629 (w), 483 (w) cm^{-1}. – HRMS (ESI, C$_{35}$H$_{26}$N$_2$) calc. 475.2169 [M+H]$^+$, found 475.2139 [M+H]$^+$.

(rac)-4-Cyano-13-(4'-N-9,10-dihydro-9,9-dimethylacridinyl)phenyl[2.2]paracyclophane (110)

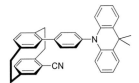

As sealable vial was charged with (*rac*)-4-bromo-13-cyano[2.2]paracyclophane (125 mg, 400 μmol, 1.00 equiv.), 4-(*N*-9,10-dihydro-9,9-dimethylacridinyl)phenyl boronic acid (198 mg, 600 μmol. 1.50 equiv.), K$_3$PO$_4$ (213 mg, 1.00 mmol, 2.50 equiv.) and Pd(PPh$_3$)$_4$ (139 mg, 120 μmol, 30 mol%). The sealed vial was evacuated and flushed with argon three times. Through the septum 7 mL of degassed toluene were added, then heated to 85 °C and stirred for 60 h. The reaction mixture was diluted with 50 mL of ethyl acetate and washed with sat. aqueous NH$_4$Cl solution (3 × 30 mL) and brine (30 mL). The organic layer was dried over Na$_2$SO$_4$ and the solvent was removed under reduced pressure. The crude product was purified by column chromatography (silica, cyclohexane/ethyl acetate; 20:1) to yield 27.5 mg of the title compound (53.2 μmol, 13%) as a white solid.

R$_f$ =0.27 (cyclohexane/ethyl acetate; 20:1). – ^{1}H NMR (400 MHz, CDCl$_3$) δ = 7.81 (d, *J* = 8.4 Hz, 2H), 7.46 (dd, *J* = 7.7, 1.6 Hz, 2H), 7.42 (d, *J* = 8.4 Hz, 2H), 7.01 (ddd, *J* = 8.4, 7.2, 1.6 Hz, 2H), 6.97 – 6.85 (m, 5H), 6.77 (dd, *J* = 7.8, 2.9 Hz, 2H), 6.59 (dd, *J* = 7.8, 1.9 Hz, 1H), 6.47 (d, *J* = 8.1 Hz, 2H), 3.85 (ddd, *J* = 13.9, 9.8, 5.8 Hz, 1H), 3.44 (ddd, *J* = 13.2, 9.8, 3.0 Hz, 1H), 3.36 – 3.21 (m, 3H), 3.17 – 2.94 (m, 3H), 1.70 (s, 6H) ppm. – ^{13}C NMR (101 MHz, CDCl$_3$) δ = 144.38 (C$_{quat}$), 142.02 (C$_{quat}$), 141.10 (C$_{quat}$), 140.76 (C$_{quat}$), 140.43 (C$_{quat}$), 139.65 (C$_{quat}$), 139.55 (C$_{quat}$), 136.94 (C$_{quat}$), 136.77 (+, C$_{Ar}$H), 136.33 (+, C$_{Ar}$H), 135.23 (+, C$_{Ar}$H), 135.14 (+, C$_{Ar}$H), 133.00 (+, C$_{Ar}$H), 132.65 (+, C$_{Ar}$H), 131.36 (+, C$_{Ar}$H), 130.92 (+, C$_{Ar}$H), 130.08 (C$_{quat}$), 126.63 (+, C$_{Ar}$H), 125.12 (+, C$_{Ar}$H), 120.61 (+, C$_{Ar}$H), 118.26 (C$_{quat,}$ CN), 114.49 (+, C$_{Ar}$H), 112.41 (C$_{quat}$), 36.13 (C$_{quat,}$ CMe$_2$), 35.26 (–, CH$_2$), 35.25 (–, CH$_2$), 34.79 (–, CH$_2$), 33.17 (–, CH$_2$), 31.27 (+, CH$_3$) ppm. – IR (ATR) $\tilde{v}$ = 2926 (w), 2215 (vw), 1739 (w), 1587 (m), 1511 (w), 1471 (m), 1448 (m), 1325 (m), 1270 (m), 1241 (m), 1111 (w), 1045 (w), 1065 (w), 913 (w), 879 (w), 844 (w), 796 (vw), 750 (m), 658 (w), 627 (w), 583 (w), 512 (w), 487 (w), 433 (w) cm^{-1}. – HRMS (APCI, C$_{38}$H$_{32}$N$_2$) calc. 517.2638 [M+H]$^+$, found 517.2622 [M+H]$^+$.

(*rac*)-4-Bromo-16-formyl[2.2]paracyclophane (112)

 In a 250 mL Schlenk flask, 4,16-dibromo[2.2]paracyclophane (3.00 g, 8.20 mmol, 1.00 equiv.) was dissolved in 200 mL of dry THF under argon atmosphere and cooled to –78 °C. Then *n*-butyllithium (3.94 mL as 2.5 M solution in hexane, 9.84 mmol, 1.20 equiv.) was added dropwise turning the reaction mixture yellow. After 1 h stirring at –78 °C, the cooling bath was removed and DMF (5.05 mL, 4.80 g, 65.6 mmol, 8.00 equiv.) were quickly *via* syringe. The reaction mixture was stirred at room temperature for 16 h and then quenched by addition of water. It was extracted with ethyl acetate (3 × 100 mL), the combined organic layers were washed with brine (100 mL), dried over Na_2SO_4 and the solvents were removed under reduced pressure. The crude product was purified by column chromatography (silica, cyclohexane/ethyl acetate) to yield 1.77 g of the title compound (5.63 mmol, 69%) as a white solid.

R_f = 0.17 (cyclohexane/ethyl acetate; 50:1). – ^{1}H NMR (400 MHz, CDCl$_3$) δ = 9.95 (s, 1H), 7.41 (dd, J = 7.7, 2.0 Hz, 1H), 6.97 (d, J = 2.0 Hz, 1H), 6.62 – 6.52 (m, 2H), 6.47 – 6.35 (m, 2H), 4.12 (ddd, J = 12.7, 9.8, 2.1 Hz, 1H), 3.50 (ddd, J = 13.3, 10.4, 2.8 Hz, 1H), 3.25 (ddd, J = 13.3, 10.4, 5.1 Hz, 1H), 3.18 – 2.99 (m, 3H), 2.99 – 2.81 (m, 2H) ppm. – ^{13}C NMR (101 MHz, CDCl$_3$) δ = 192.22 (+, CHO), 142.84 (C$_{quat}$), 141.59 (C$_{quat}$), 140.23 (C$_{quat}$), 139.05 (C$_{quat}$), 137.32 (+, C$_{Ar}$H), 137.14 (+, C$_{Ar}$H), 136.68 (C$_{quat}$), 135.28 (+, C$_{Ar}$H), 134.16 (+, C$_{Ar}$H), 133.87 (+, C$_{Ar}$H), 130.82 (+, C$_{Ar}$H), 127.47 (C$_{quat}$), 35.29 (–, CH$_2$), 34.27 (–, CH$_2$), 33.27 (–, CH$_2$), 33.07 (–, CH$_2$) ppm. – IR (ATR) $\tilde{v}$ = 2925 (w), 2851 (w), 1678 (m), 1584 (m), 1548 (w), 1474 (w), 1433 (w), 1390 (w), 1282 (w), 1224 (w), 1194 (w), 1142 (w), 1029 (w), 970 (w), 892 (w), 872 (w), 836 (w), 739 (w), 708 (w), 647 (m), 622 (w), 522 (w), 496 (w), 476 (w), 452 (w), 400 (w) cm^{-1}. – HRMS (APCI, C$_{17}$H$_{15}$BrO) calc. 315.0,79 [M(^{81}Br)+H]$^+$, found 315.0364 [M(^{81}Br)]$^+$; calc. 317.0359 [M(^{81}Br)+H]$^+$, found 317.0324 [M(^{81}Br)+H]$^+$. – EA (C$_{17}$H$_{15}$BrO) calc. C:64.78, H: 4.80; found C: 64.87, H: 4.93. The analytical data is consistent with literature.[224]

(*rac*)-4-Bromo-16-cyano[2.2]paracyclophane (113)

 In a 250 mL round-bottom flask, (*rac*)-4-bromo-16-formyl[2.2]paracyclophane (1.86 g, 5.90 mmol, 1.00 equiv.) and hydroxylammonium chloride (4.07 g, 59.0 mmol, 10.00 equiv.) were dissolved in 150 mL of dimethylsulfoxide and stirred at 100 °C for 2 h. After cooling to room temperature, the reaction mixture was diluted with sat. NaHCO₃ solution and extracted with ethyl acetate (3 × 80 mL), then the combined organic layers were thoroughly washed with brine (3 × 150 mL) to remove DMSO. The organic layers were dried over Na₂SO₄ and the solvent was removed under reduced pressure. The crude product was purified *via* column chromatography (silica, cyclohexane/ethyl acetate; 20:1) to yield 1.10 g of the title compound (3.54 mmol, 60%) as a white solid.

R_f = 0.28 (cyclohexane/ethyl acetate; 20:1). – ¹H NMR (400 MHz, CDCl₃) δ = 7.39 (dd, J = 8.0, 1.9 Hz, 1H), 6.93 (dd, J = 7.9, 1.9 Hz, 1H), 6.75 (d, J = 1.9 Hz, 1H), 6.57 (d, J = 8.0 Hz, 1H), 6.54 (d, J = 1.9 Hz, 1H), 6.49 (d, J = 7.9 Hz, 1H), 3.58 – 3.44 (m, 2H), 3.31 – 3.17 (m, 2H), 3.12 – 2.94 (m, 3H), 2.87 (ddd, J = 13.5, 10.6, 5.7 Hz, 1H) ppm. – ¹³C NMR (101 MHz, CDCl₃) δ = 143.82 (C$_{quat}$), 141.11 (C$_{quat}$), 140.68 (C$_{quat}$), 139.11 (C$_{quat}$), 137.74 (+, C$_{Ar}$H), 137.26 (+, C$_{Ar}$H), 134.65 (+, C$_{Ar}$H), 133.73 (+, C$_{Ar}$H), 133.11 (+, C$_{Ar}$H), 129.83 (+, C$_{Ar}$H), 127.12 (C$_{quat}$), 119.01 (C$_{quat}$, CN), 115.04 (C$_{quat}$), 35.56 (–, CH₂), 33.98 (–, CH₂), 33.82 (–, CH₂), 33.01 (–, CH₂) ppm. – IR (ATR) $\tilde{v}$ = 2932 (w), 2851 (vw), 2217 (w), 1583 (vw), 1539 (vw), 1474 (w), 1447 (w), 1388 (w), 1187 (vw), 1154 (vw), 1032 (w), 952 (vw), 905 (w), 891 (w), 870 (w), 836 (w), 709 (w), 669 (w), 651 (w), 614 (vw), 500 (w), 475 (w), 402 (vw) cm⁻¹. – HRMS (ESI, C₁₇H₁₄BrN) calc. 312.0382 [M(⁷⁹Br)+H]⁺, found 312.0375 [M(⁷⁹Br)+H]⁺; calc 314.0362 [M(⁸¹Br)+H]⁺, found 314.0353 [M(⁸¹Br)+H]⁺. – EA (C₁₇H₁₄BrN) calc. C: 65.40, H: 4.52, N: 4.49; found C: 65.33, H: 4.63, N: 4.41.

(*rac*)-4-Cyano-16-(4'-*N,N*-diphenylamino)phenyl[2.2]paracyclophane (114)

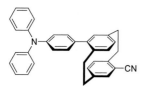

As sealable vial was charged with (*rac*)-4-bromo-16-cyano[2.2]paracyclophane (125 mg, 400 µmol, 1.00 equiv.), 4-(*N,N*-diphenylamino)phenyl boronic acid (231 mg, 800 µmol, 2.00 equiv.), K_3PO_4 (170 mg, 800 µmol, 2.00 equiv.), $Pd(OAc)_2$ (4.50 mg, 20.0 µmol, 5 mol%) and 2-dicyclohexylphosphino-2,6-diisopropoxybiphenyl (18.7 mg, 40.0 µmol, 10 mol%). The sealed vial was evacuated and purged with argon three times. Through the septum 7 mL of degassed toluene and 1 mL of degassed water were added, then heated to 75 °C and stirred for 16 h. The reaction mixture was diluted with 50 mL of ethyl acetate and washed with sat. aqueous NH_4Cl solution (3 × 30 mL) and brine (30 mL). The organic layer was dried over Na_2SO_4 and the solvent was removed under reduced pressure. The crude product was purified by column chromatography (silica, cyclohexane/ethyl acetate; 20:1) to yield 177 mg of the title compound (371 µmol, 93%) as a white solid.

R_f = 0.19 (cyclohexane/ethyl acetate; 20:1). – ^{1}H NMR (400 MHz, $CDCl_3$) δ = 7.34 – 7.27 (m, 6H), 7.20 – 7.14 (m, 6H), 7.07 (t, J = 7.3 Hz, 2H), 6.90 – 6.83 (m, 3H), 6.65 – 6.55 (m, 3H), 3.63 – 3.46 (m, 2H), 3.34 (t, J = 9.7 Hz, 1H), 3.18 – 3.05 (m, 2H), 3.03 – 2.75 (m, 3H) ppm. – ^{13}C NMR (101 MHz, $CDCl_3$) δ = 147.72 (C_{quat}), 147.10 (C_{quat}), 144.27 (C_{quat}), 142.04 (C_{quat}), 141.30 (C_{quat}), 139.25 (C_{quat}), 137.06 (+, $C_{Ar}H$), 136.90 (C_{quat}), 135.51(+, $C_{Ar}H$), 134.74 (C_{quat}), 134.27 (+, $C_{Ar}H$), 133.58 (+, $C_{Ar}H$), 132.42 (+, $C_{Ar}H$), 130.34 (+, $C_{Ar}H$), 130.00 (+, $C_{Ar}H$), 129.49 (+, $C_{Ar}H$), 124.84 (+, $C_{Ar}H$), 123.31 (+, $C_{Ar}H$), 123.27 (+, $C_{Ar}H$), 119.22 (C_{quat}, CN), 114.75 (C_{quat}), 34.60 (–, CH_2), 34.44 (–, CH_2), 34.17 (–, CH_2), 34.05 (–, CH_2) ppm. – IR (ATR) $\tilde{v}$ = 2926 (vw), 2217 (vw), 1587 (w), 1509 (w), 1488 (w), 1317 (w), 1270 (w), 1026 (vw), 872 (w), 838 (w), 771 (vw), 745 (w), 693 (m), 659 (w), 616 (w), 567 (vw), 510 (w), 500 (w), 444 (vw), 405 (vw) cm^{-1}. – HRMS (ESI, $C_{35}H_{28}N_2$) calc. 476.2252 $[M]^+$, found 476.2239 $[M]^+$. – EA ($C_{35}H_{28}N_2$) calc. C 88.20, H: 5.92, N: 5.88; found C: 88.21, H: 6.15, N: 5.69.

(rac)-4-Cyano-16-(4'-*N*-carbazolyl)phenyl[2.2]paracyclophane (115)

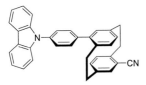 A sealable vial was charged with *(rac)*-4-bromo-16-cyano[2.2]paracyclophane (125 mg, 400 μmol, 1.00 equiv.), 4-(*N*-carbazolyl)phenyl boronic acid (230 mg, 800 μmol, 2.00 equiv.), K$_3$PO$_4$ (170 mg, 800 μmol, 2.00 equiv.), Pd(OAc)$_2$ (4.50 mg, 20.0 μmol, 5 mol%) and 2-dicyclohexylphosphino-2,6-diisopropoxybiphenyl (18.7 mg, 40.0 μmol, 10 mol%). The sealed vial was evacuated and purged with argon three times. Through the septum 7 mL of degassed toluene and 1 mL of degassed water were added, then heated to 75 °C and stirred for 16 h. The reaction mixture was diluted with 50 mL of ethyl acetate and washed with sat. aqueous NH$_4$Cl solution (3 × 30 mL) and brine (30 mL). The organic layer was dried over Na$_2$SO$_4$ and the solvent was removed under reduced pressure. The crude product was purified by column chromatography (silica, cyclohexane/ethyl acetate; 20:1) to yield 143 mg of the title compound (301 μmol, 75%) as a white solid.

R$_f$ = 0.21 (cyclohexane/ethyl acetate; 20:1). – ^{1}H NMR (400 MHz, CDCl$_3$) δ = 8.19 (d, *J* = 7.8 Hz, 2H), 7.73 – 7.65 (m, 4H), 7.55 (d, *J* = 8.2 Hz, 2H), 7.47 (ddd, *J* = 8.2, 7.1, 1.2 Hz, 2H), 7.34 (ddd, *J* = 8.2, 7.1, 1.2 Hz, 2H), 6.97 (dd, *J* = 7.8, 2.0 Hz, 1H), 6.93 – 6.89 (m, 2H), 6.76 – 6.71 (m, 2H), 6.67 (d, *J* = 7.8 Hz, 1H), 3.75 – 3.51 (m, 2H), 3.40 (t, *J* = 9.9 Hz, 1H), 3.26 – 3.05 (m, 3H), 3.03 – 2.79 (m, 2H) ppm. – ^{13}C NMR (101 MHz, CDCl$_3$) δ = 144.39 (C$_{quat}$), 141.51 (C$_{quat}$), 141.28 (C$_{quat}$), 140.87 (C$_{quat}$), 139.92 (C$_{quat}$), 139.59 (C$_{quat}$), 137.16 (+, C$_{Ar}$H), 137.13 (C$_{quat}$), 136.90 (C$_{quat}$), 135.71 (+, C$_{Ar}$H), 134.29 (+, C$_{Ar}$H), 133.68 (+, C$_{Ar}$H), 132.66 (+, C$_{Ar}$H), 131.02 (+, C$_{Ar}$H), 130.75 (+, C$_{Ar}$H), 127.28 (+, C$_{Ar}$H), 126.13 (+, C$_{Ar}$H), 123.63 (C$_{quat}$), 120.54 (+, C$_{Ar}$H), 120.24 (+, C$_{Ar}$H), 119.14 (C$_{quat}$, CN), 114.89 (C$_{quat}$), 109.95 (+, C$_{Ar}$H), 34.75 (–, CH$_2$), 34.46 (–, CH$_2$), 34.21 (–, CH$_2$), 33.88 (–, CH$_2$) ppm. – IR (ATR) $\tilde{v}$ = 2923 (vw), 2218 (w), 1598 (w), 1512 (m), 1479 (w), 1450 (m), 1361 (w), 1337 (w), 1317 (w), 1230 (m), 1107 (w), 1017 (w), 913 (w), 875 (w), 842 (w), 751 (s), 726 (m), 653 (w), 625 (w), 565 (w), 505 (w), 428 (w) cm^{-1}. – HRMS (ESI, C$_{35}$H$_{26}$N$_2$) calc. 474.2096 [M]$^+$, found 474.2082 [M]$^+$. – EA (C$_{35}$H$_{26}$N$_2$) calc. C: 88.58, H: 5.52, N: 5.90; found C: 88.56, H: 5.72, N: 5.83.

(rac)-4-Cyano-16-(4'-*N*-9,10-dihydro-9,9-dimethylacridinyl)phenyl[2.2]paracyclophane (116)

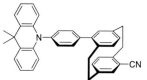

A sealable vial was charged with *(rac)*-4-bromo-16-cyano[2.2]paracyclophane (125 mg, 400 µmol, 1.00 equiv.), 4-(*N*-9,10-dihydro-9,9-dimethylacridinyl)phenyl boronic acid (263 mg, 800 µmol, 2.00 equiv.), K₃PO₄ (170 mg, 800 µmol, 2.00 equiv.), Pd(OAc)₂ (4.50 mg, 20.0 µmol, 5 mol%) and 2-dicyclohexylphosphino-2,6-diisopropoxybiphenyl (18.7 mg, 40.0 µmol, 10 mol%). The sealed vial was evacuated and purged with argon three times. Through the septum 7 mL of degassed toluene and 1 mL of degassed water were added, then heated to 75 °C and stirred for 16 h. The reaction mixture was diluted with 50 mL of ethyl acetate and washed with sat. aqueous NH₄Cl solution (3 × 30 mL) and brine (30 mL). The organic layer was dried over Na₂SO₄ and the solvent was removed under reduced pressure. The crude product was purified by column chromatography (silica, cyclohexane/ethyl acetate; 20:1) to yield 149 mg of the title compound (288 µmol, 72%) as a white solid.

R_f = 0.23 (cyclohexane/ethyl acetate; 20:1). – ¹H NMR (400 MHz, CDCl₃) δ = 7.71 (d, J = 8.3 Hz, 2H), 7.50 (dd, J = 7.7, 1.7 Hz, 2H), 7.47 (d, J = 8.3 Hz, 2H), 7.04 (ddd, J = 8.2, 7.2, 1.7 Hz, 2H), 7.00 – 6.90 (m, 5H), 6.76 – 6.71 (m, 2H), 6.68 (d, J = 7.8 Hz, 1H), 6.39 (dd, J = 8.2, 1.4 Hz, 2H), 3.69 – 3.54 (m, 2H), 3.46 – 3.36 (m, 1H), 3.26 – 3.04 (m, 3H), 2.99 (ddd, J = 13.5, 10.0, 3.7 Hz, 1H), 2.85 (ddd, J = 13.5, 9.9, 4.8 Hz, 1H), 1.74 (s, 6H) ppm. – ¹³C NMR (101 MHz, CDCl₃) δ = 144.37 (C_quat), 141.48 (C_quat), 141.20 (C_quat), 141.05 (C_quat), 140.77 (C_quat), 140.34 (C_quat), 139.52 (C_quat), 137.13 (+, C_ArH), 137.12 (C_quat), 135.71 (+, C_ArH), 134.28 (+, C_ArH), 133.62 (+, C_ArH), 132.77 (+, C_ArH), 131.97 (+, C_ArH), 131.72 (+, C_ArH), 130.70 (+, C_ArH), 130.30 (C_quat), 126.50 (+, C_ArH), 125.38 (+, C_ArH), 120.82 (+, C_ArH), 119.14 (C_quat, CN), 114.93 (C_quat), 114.11 (+, C_ArH), 36.15 (C_quat, CMe₂), 34.67 (–, CH₂), 34.46 (–, CH₂), 34.16 (–, CH₂), 34.06 (–, CH₂), 31.25 (+, CH₃) ppm. – IR (ATR) $\tilde{v}$ = 2923 (w), 2853 (w), 2220 (w), 1587 (w), 1509 (w), 1469 (m), 1445 (m), 1324 (m), 1268 (m), 1043 (w), 1016 (w), 925 (w), 871 (w), 842 (w), 796 (w), 746 (m), 670 (w), 654 (w), 628 (w), 601 (w), 581 (w), 559 (w), 501 (m), 436 (w) cm⁻¹. – HRMS (APCI, C₃₈H₃₂N₂) calc. 517.2638 [M+H]⁺, found 517.2625 [M+H]⁺.

5.3   Analytical Data of Carbazolophane Donor for TADF Turn-On Project

(rac)-4-Bromo[2.2]paracyclophane (17)

A solution of bromine (3.77 mL, 11.8 g, 73.7 mmol, 1.02 equiv.) in 80 mL of dichloromethane was prepared in a dropping funnel and 5 mL of this solution were added to iron filings (10.0 mg, 2 mol%, 1.44 mmol) in a 500 mL three-necked flask and stirred for 1 h at room temperature. Then 250 mL of dichloromethane and [2.2]paracyclophane (15.0 g, 72.0 mmol, 1.00 equiv.) was added and stirred for another 30 min. The remaining bromine solution was added dropwise over a period of 5 h and the mixture was stirred for 3 days. The reaction was quenched with saturated Na_2SO_3 solution (200 mL) and stirred for 30 min until full discoloration of the mixture. The organic phase was separated, washed with brine (200 mL) and dried over Na_2SO_4. The solvent was removed under reduced pressure to yield 20.2 g of the title compound (70.3 mmol, 98%) as a white solid.

R_f = 0.63 (cyclohexane/ethyl acetate; 50:1). – ^{1}H NMR (400 MHz, CDCl$_3$) δ = 7.19 (dd, J = 7.9, 2.0 Hz, 1H), 6.58 (dd, J = 7.8, 2.0 Hz, 1H), 6.56 – 6.52 (m, 2H), 6.52 – 6.45 (m, 3H), 3.48 (ddd, J = 13.0, 10.1, 2.2 Hz, 1H), 3.22 (ddd, J = 13.1, 10.1, 6.1 Hz, 1H), 3.17 – 3.02 (m, 4H), 2.98 – 2.78 (m, 2H) ppm. – ^{13}C NMR (101 MHz, CDCl$_3$) δ = 141.73 (C_{quat}), 139.43 (C_{quat}), 139.22 (C_{quat}, 2C), 137.37 (+, $C_{Ar}H$), 135.17 (+, $C_{Ar}H$), 133.43 (+, $C_{Ar}H$), 133.03 (+, $C_{Ar}H$), 132.37 (+, $C_{Ar}H$), 131.59 (+, $C_{Ar}H$), 128.81 (+, $C_{Ar}H$), 127.09 (C_{quat}), 35.98 (–, CH_2), 35.61 (–, CH_2), 34.95 (–, CH_2), 33.61 (–, CH_2) ppm. – IR (ATR) $\tilde{v}$ = 2924 (w), 2849 (w), 1585 (w), 1541 (w), 1497 (w), 1475 (w), 1431 (w), 1408 (w), 1390 (w), 1186 (w), 1092 (vw), 1034 (m), 941 (w), 869 (w), 839 (m), 793 (w), 708 (m), 668 (w), 640 (m), 576 (w), 514 (m), 473 (w), 404 (vw), 382 (w) cm^{-1}. – MS (EI, 70 eV) m/z (%) = 288 (27) [M(^{81}Br)]$^+$, 286 (28) [M(^{79}Br)]$^+$, 184 (17) [C_8H_7^{81}Br]$^+$, 182 (17) [C_8H_7^{79}Br], 104 (100) [C_8H_8]$^{+\cdot}$. – HRMS (EI, $C_{16}H_{15}$^{79}Br) calc. 286.0352; found 286.0350. The analytical data is consistent with literature.[225]

(rac)-4-*N*-(Phenyl)amino[2.2]paracyclophane (36)

 As sealable vial was charged with *(rac)*-4-bromo[2.2]paracyclophane (1.15 g, 4.00 mmol, 1.00 equiv.), potassium *tert*-butoxide (630 mg, 5.60 mmol, 1.40 equiv.), Pd2dba3 (70.0 mg, 0.08 mmol, 2 mol%) and SPhos (100 mg, 0.240 µmol, 6 mol%) and evacuated and purged with argon three times. Through the septrum, 10 mL of dry toluene and aniline (440 µL, 450 mg, 4.80 mmol, 1.20 equiv.) were added. The mixture was stirred at 115 °C for 16 h. The crude product was diluted with water (100 mL) and extracted with ethyl acetate (3 × 200 mL), then washed with saturated NaHCO3 solution (100 mL) and brine (100 mL). The combined organic layers were dried over Na2SO4 and the solvent was evaporated under reduced pressure. The residue was purified by column chromatography (silica, cyclohexane/ethyl acetate; 50:1) to yield 988 mg of the title compound (3.30 mmol, 82%) as a white solid.

R_f = 0.35 (cyclohexane/ethyl acetate; 50:1). – ^{1}H NMR (500 MHz, CDCl3) δ = 7.23 (t, J = 7.9 Hz, 2H), 7.01 (dd, J = 7.8, 1.9 Hz, 1H), 6.93 (d, J = 8.5 Hz, 2H), 6.87 (t, J = 7.3 Hz, 1H), 6.59 (dd, J = 7.8, 1.9 Hz, 1H), 6.51 – 6.41 (m, 3H), 6.37 (dd, J = 7.8, 1.8 Hz, 1H), 5.86 (d, J = 1.8 Hz, 1H), 5.55 (bs, 1H), 3.12 – 2.86 (m, 7H), 2.69 (ddd, J = 13.7, 10.1, 6.8 Hz, 1H) ppm. – ^{13}C NMR (126 MHz, CDCl3) δ = 143.74 (C_{quat}), 141.35 (C_{quat}), 140.21 (C_{quat}), 139.64 (C_{quat}), 139.13 (C_{quat}), 135.99 (+, C_{Ar}H), 133.64 (+, C_{Ar}H), 132.85 (+, C_{Ar}H), 131.48 (C_{quat}), 131.34 (+, C_{Ar}H), 129.28 (+, C_{Ar}H), 127.55 (+, C_{Ar}H), 127.04 (+, C_{Ar}H), 125.99 (+, C_{Ar}H), 120.17 (+, C_{Ar}H), 116.09 (+, C_{Ar}H), 35.35 (–, CH2), 35.01 (–, CH2), 33.95 (–, CH2), 33.93 (–, CH2) ppm. – IR (ATR) ṽ = 3372 (w, ν(NH)), 3005 (vw), 2922 (0.85), 2851 (vw), 1591 (w), 1563 (w), 1492 (m), 1433 (w), 1305 (w), 1280 (w), 1239 (w), 1173 (w), 1151 (w), 1108 (w), 1091 (w), 1027 (vw), 996 (vw), 977 (vw), 938 (vw), 886 (w), 800 (w), 780 (vw), 741 (m), 716 (w), 693 (m), 659 (w), 642 (w), 615 (vw), 588 (vw) cm^{-1}. – MS (EI, 70 eV) *m/z* (%) = 300 (17) [M+H]$^+$, 299 (73) [M]$^+$, 298 (34) [M–H]$^+$, 195 (78) [M–C8H8]$^+$, 194 (84) [M–C8H9]$^+$, 193 (24) [M–C8H10]$^+$, 104 (100) [C8H8]$^+$. – HRMS (EI, C22H21N) calc. 299.1669; found 299.1670. – HRMS (APCI, C22H21N) calc. 300.1747 [M+H]$^+$; found 300.1737 [M+H]$^+$. The analytical data is consistent with literature.[39]

(*rac*)-[2]Paracyclo[2](1,4)carbazolophane (Carbazolophane) (37)

In a 500 mL three-necked flask equipped with a reflux condenser, (*rac*)-4-*N*-(phenyl)amino[2.2]paracyclophane (8.08 g, 27.0 mmol, 1.00 equiv.), palladium acetate (31.52 g, 6.75 mmol, 25 mol%) and pivalic acid (2.07 g, 20.3 mmol, 75 mol%) were dissolved in 200 mL of acetic acid and the mixture was stirred at 120 °C for 7 h. While stirring, compressed air was added continuously to the suspension using a Teflon hose. The crude product was dissolved in water (300 mL) and neutralized by adding solid potassium hydroxide and solid sodium bicarbonate. The mixture was extracted with ethyl acetate (3 × 300 mL) and the combined organic layers were washed with brine (400 mL) and dried over Na_2SO_4. The dark brown filtrate was evaporated under reduced pressure and the residue was purified by column chromatography (silica, cyclohexane/ethyl acetate; 10:1) to yield 2.07 g of the title compound (6.78 mmol, 26%) as a white solid.

R_f = 0.29 (cyclohexane/ethyl acetate; 10:1). – ¹H NMR (400 MHz, CDCl₃) δ = 8.06 (d, *J* = 7.9 Hz, 1H), 7.88 (bs, NH, 1H), 7.47 (d, *J* = 8.0 Hz, 1H), 7.40 (t, *J* = 7.6 Hz, 1H), 7.32 – 7.20 (m, 1H), 6.63 (d, *J* = 7.5 Hz, 1H), 6.56 (d, *J* = 7.5 Hz, 1H), 6.50 (dd, *J* = 7.8, 1.8 Hz, 1H), 6.36 (dd, *J* = 7.8, 1.9 Hz, 1H), 5.94 (dd, *J* = 7.8, 1.9 Hz, 1H), 5.22 (dd, *J* = 7.8, 1.8 Hz, 1H), 4.03 –3.98 (m, 1H), 3.44 – 3.26 (m, 1H), 3.19 – 2.99 (m, 5H), 2.94 (ddd, *J* = 12.0, 9.9, 6.1 Hz, 1H). – ¹³C NMR (101 MHz, CDCl₃) δ = 140.14 (C_{quat}), 138.94 (C_{quat}), 138.01 (C_{quat}), 137.41 (C_{quat}), 135.89 (C_{quat}), 132.12 (+, $C_{Ar}H$), 131.60 (+, $C_{Ar}H$), 131.09 (+, $C_{Ar}H$), 126.57 (+, $C_{Ar}H$), 126.41 (+, $C_{Ar}H$), 125.52 (C_{quat}), 125.35 (C_{quat}), 124.97 (+, $C_{Ar}H$), 124.67 (+, $C_{Ar}H$), 122.53 (+, $C_{Ar}H$), 122.31 (C_{quat}), 119.78 (+, $C_{Ar}H$), 110.76 (+, $C_{Ar}H$), 34.05 (–, CH_2), 33.86 (–, CH_2), 33.35 (–, CH_2), 31.22 (–, CH_2) ppm. – IR (ATR) $\tilde{v}$ = 3389 (w, *v*(NH)), 3031 (vw), 3005 (vw), 2986 (vw), 2923 (w), 2581 (vw), 1592 (vw), 1499 (w), 1455 (w), 1436 (w), 1393 (w), 1321 (w), 1298 (w), 1246 (w), 1026 (w), 931 (w), 871 (w), 805 (w), 770 (vw), 749 (m), 735 (m), 719 (w), 671 (w), 609 (w), 585 (w), 569 (vw), 514 (w), 470 (w) cm⁻¹. – HRMS (APCI, $C_{22}H_{19}N$) calc. 298.1590 [M+H]⁺; found 298.1581 [M+H]⁺. The analytical data is consistent with literature.[39]

(rac)-1-(N-[2]Paracyclo[2](1,4)carbazolophanyl)-4-cyanobenzene (125)

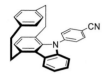

In a sealable vial, (rac)-[2]paracyclo[2](1,4)carbazolophane (20.0 mg, 67.0 µmol, 1.00 equiv.), 4-fluorobenzonitrile (54.5 mg, 450 µmol, 6.71 equiv.) and tripotassium phosphate (130 mg, 600 µmol, 8.96 equiv.) were dissolved in 5 mL of DMSO and stirred at 115 °C for 16 h. After cooling to room temperature, the mixture was diluted with 30 mL of ethyl acetate and washed with brine (3 × 50 mL). The organic layer was dried over Na_2SO_4. Afterwards, the solvent was evaporated under reduced pressure and the residue was purified by column chromatography (silica, cyclohexane/ethyl acetate; 20:1) to yield 18.4 mg of the title compound (46.2 µmol, 69%) as a white and luminescent solid.

R_f = 0.28 (cyclohexane/ethyl acetate; 20:1). – ^{1}H NMR (500 MHz, CDCl$_3$) δ = 8.14 (dd, J = 7.0, 1.4 Hz, 1H), 7.89 (d, J = 8.1 Hz, 2H), 7.70 (bs, 2H), 7.45 – 7.34 (m, 3H), 6.67 (s, 2H), 6.50 (dd, J = 7.8, 1.9 Hz, 1H), 6.34 (dd, J = 7.8, 1.9 Hz, 1H), 5.86 (dd, J = 7.7, 1.9 Hz, 1H), 5.49 (dd, J = 7.7, 1.9 Hz, 1H), 4.14 – 3.98 (m, 1H), 3.28 – 3.01 (m, 3H), 2.97 – 2.87 (m, 1H), 2.74 (ddd, J = 14.2, 10.2, 7.5 Hz, 1H), 2.59 (ddd, J = 14.2, 9.6, 1.6 Hz, 1H), 2.16 (ddd, J = 13.3, 9.6, 7.5 Hz, 1H) ppm. – ^{13}C NMR (126 MHz, CDCl$_3$) δ = 143.26 (C$_{quat}$), 140.25 (C$_{quat}$), 140.10 (C$_{quat}$), 137.73 (C$_{quat}$), 137.60 (C$_{quat}$), 136.14 (C$_{quat}$), 133.91 (+, C$_{Ar}$H), 133.57 (+, C$_{Ar}$H), 131.85 (+, C$_{Ar}$H), 131.41 (+, C$_{Ar}$H), 128.39 (+, C$_{Ar}$H), 127.48 (C$_{quat}$), 127.42 (+, C$_{Ar}$H), 126.70 (+, C$_{Ar}$H), 126.13 (C$_{quat}$), 125.71 (+, C$_{Ar}$H), 125.59 (+, C$_{Ar}$H), 124.29 (C$_{quat}$), 122.84 (+, C$_{Ar}$H), 121.60 (+, C$_{Ar}$H), 118.51 (C$_{quat}$, CN), 110.50 (C$_{quat}$), 109.42 (+, C$_{Ar}$H), 35.02 (–, CH$_2$), 33.97 (–, CH$_2$), 33.47 (–, CH$_2$), 33.17 (–, CH$_2$) ppm. – IR (ATR) ṽ = 3505 (vw), 3415 (vw), 3298 (vw), 3231 (vw), 3024 (vw), 2924 (vw), 2851 (vw), 2683 (vw), 2518 (vw), 2324 (vw), 2226 (vw), 2066 (vw), 1874 (vw), 1736 (vw), 1597 (w), 1452 (w), 1290 (vw), 1173 (vw), 1024 (vw), 890 (vw), 748 (w), 555 (w), 433 (vw) cm^{-1}.– HRMS (ESI, C$_{29}$H$_{22}$N$_2$) calc. 399.1856 [M+H]$^+$; found 399.1837 [M+H]$^+$.

2-(4-Fluorophenyl)-4,6-diphenyl-1,3,5-triazine (119)

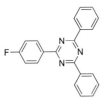 In a 250 mL Schlenk flask equipped with a reflux condenser, 2-chlorophenyl-4,6-diphenyl-1,3,5-triazine (5.17 g, 19.3 mmol, 1.00 equiv.), (4-fluorophenyl)boronic acid (4.06 g, 29.0 mmol, 1.50 equiv.), potassium carbonate (5.34 g, 38.7 mmol, 2.00 equiv.) and Pd(PPh$_3$)$_4$ (1.50 g, 1.30 mmol, 7 mol%) were dissolved in 150 mL of dry toluene under argon atmosphere. Deuterium oxide (0.50 mL) was added. The slightly yellow reaction mixture was stirred at 75 °C for 19 h. The mixture was then diluted with 100 mL of water and 100 mL of ethyl acetate. The layers were separated and the aqueous layer was extracted with ethyl acetate (3 × 50 mL). The combined organic layers were washed with brine (3 × 100 mL) and dried over Na$_2$SO$_4$. The yellow filtrate was evaporated under reduced pressure and the crude product was purified by column chromatography (silica, cyclohexane/ethyl acetate; 2:1) to yield 4.55 g of the title compound (13.9 mmol, 72%) as a white fibrous solid.

R$_f$ = 0.50 (cyclohexane/dichloromethane; 50:1). – ^{1}H NMR (500 MHz, CDCl$_3$) δ = 8.83 – 8.71 (m, 6H), 7.66 – 7.54 (m, 6H), 7.25 (t, J = 8.6 Hz, 2H) ppm. – ^{13}C NMR (125 MHz, CDCl$_3$) δ = 171.75 (C$_{quat}$), 170.76 (C$_{quat}$), 165.93 (C$_{quat}$, d, $^1J_{CF}$ = 253.0 Hz), 136.23 (C$_{quat}$), 132.73 (+, C$_{Ar}$), 132.52 (C$_{quat}$, d, $^4J_{CF}$ = 2.9 Hz), 131.43 (+, d, $^3J_{CF}$ = 9.0 Hz, C$_{Ar}$H), 129.08 (+, C$_{Ar}$), 128.79 (+, C$_{Ar}$), 115.84 (+, d, $^2J_{CF}$ = 21.8 Hz, C$_{Ar}$H) ppm. – ^{19}F NMR (470 MHz, CDCl$_3$) δ = –107.1 ppm. – IR (ATR) $\tilde{v}$ = 3055 (w), 1915 (w), 1601 (m), 1509 (s), 1446 (m), 1412 (m), 1362 (s), 1224 (m), 1175 (m), 1142 (m), 1066 (m), 1026 (m), 855 (m), 832 (m), 764 (s), 728 (m), 685 (s), 644 (m), 583 (m), 503 (m), 469 (w), 392 (w) cm^{-1}.– EA (C$_{21}$H$_{14}$FN$_3$) calc. N: 12.84, C:77.05, H: 4.31; found N: 12.80, C: 76.95, H: 4.31. – HRMS (APCI, C$_{21}$H$_{14}$FN$_3$) calc. 328.1245 [M+H]$^+$; found 328.1234 [M+H]$^+$. The analytical data is consistent with literature.[226]

9-(4-(4,6-Diphenyl-1,3,5-triazin-2-yl)phenyl)-9*N*-carbazole (76)

In a sealable vial, carbazole (669 mg, 4.00 mmol, 3.08 equiv.), 2-(4-fluorophenyl)-4,6-diphenyl-1,3,5-triazine (426 mg, 1.30 mmol, 1.00 equiv.) and tripotassium phosphate (1.27 g, 6.00 mmol, 4.62 equiv.) were dissolved in 8 mL of DMSO and stirred at 115 °C for 16 h. After cooling to room temperature, the mixture was diluted with ethyl acetate (100 mL) and washed with brine (3 × 100 mL) to remove DMSO. The organic layer was dried over Na$_2$SO$_4$ and the solvent was removed under reduced pressure. The crude solid was purified by column chromatography (silica, gradient cyclohexane/dichloromethane; 50:1 to 10:1 to 5:1 to 2:1) to yield 386 mg of the title compound (813 μmol, 49%) as a yellow luminescent solid.

R$_f$ = 0.28 (cyclohexane/dichloromethane; 10:1). – Mp = 270 °C. – ^{1}H NMR (500 MHz, CDCl$_3$) δ = 9.03 (d, J = 8.6 Hz, 2H), 8.83 (d, J = 8.2 Hz, 4H), 8.18 (d, J = 7.8 Hz, 2H), 7.83 (d, J = 8.5 Hz, 2H), 7.69 – 7.59 (m, 6H), 7.57 (d, J = 8.2 Hz, 2H), 7.46 (t, J = 7.7 Hz, 2H), 7.34 (t, J = 7.5 Hz, 2H) ppm. – ^{13}C NMR (126 MHz, CDCl$_3$) δ = 171.95 (C$_{quat}$), 171.08 (C$_{quat}$), 141.75 (C$_{quat}$), 140.57 (C$_{quat}$), 136.27 (C$_{quat}$), 135.09 (C$_{quat}$), 132.82 (+, C$_{Ar}$H), 130.79 (+, C$_{Ar}$H), 129.16 (+, C$_{Ar}$H), 128.87(+, C$_{Ar}$H), 126.88 (+, C$_{Ar}$H), 126.30 (+, C$_{Ar}$H), 123.89 (C$_{quat}$), 120.58 (+, C$_{Ar}$H), 120.55 (+, C$_{Ar}$H), 110.07 (+, C$_{Ar}$H) ppm. – IR (ATR) $\tilde{v}$ = 3054 (vw), 2850 (vw), 2921 (w), 1589 (w), 1512 (m), 1444 (m), 1366 (m), 1223 (w), 1169 (w), 1026 (w), 827 (w), 765 (m), 737 (m), 720 (m), 686 (m), 647 (w), 621 (w), 562 (w), 505 (w), 466 (w), 418 (w) cm^{-1}. – EA (C$_{33}$H$_{22}$N$_4$) calc. C: 83.52, H: 4.67, N: 11.81; found C: 82.29, H: 4.56, N: 11.54. – HRMS (APCI, C$_{33}$H$_{22}$N$_4$) calc. 475.1917 [M+H]$^+$; found 475.1900 [M+H]$^+$. The analytical data is consistent with literature.[227]

(rac)-1-(N-[2]Paracyclo[2](1,4)carbazolophanyl)-4-(4,6-diphenyl-1,3,5-triazin-2-yl)-benzene (117)

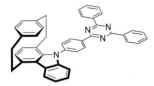

In a sealable vial, *(rac)*-[2]paracyclo[2](1,4)carbazolophane (119 mg, 400 µmol, 1.00 equiv.), 2-(4-fluorophenyl)-4,6-diphenyl-1,3,5-triazine (196 mg, 600 µmol, 1.50 equiv.) and tripotassium phosphate (187 mg, 880 µmol, 2.20 equiv.) were dissolved in 10 mL of DMSO and stirred at 115 °C for 16 h. After cooling to room temperature, the mixture was diluted with ethyl acetate (100 mL) and washed with brine (3 × 100 mL) to remove DMSO. The organic layer was dried over Na$_2$SO$_4$ and the solvent was removed under reduced pressure. The crude solid was purified by column chromatography (silica, cyclohexane/dichloromethane; 100:1 to 50:1 to 10:1 to 5:1 to 1:1) to yield 127 mg of the title compound (210 µmol, 52%) as a yellow and luminescent solid.

R_f = 0.28 (cyclohexane/dichloromethane; 10:1). – Mp = 263 °C. – ^{1}H NMR (500 MHz, CDCl$_3$) δ = 9.03 (bs, 2H), 8.88 – 8.81 (m, 4H), 8.17 (d, J = 7.8 Hz, 1H), 7.72 – 7.57 (m, 8H), 7.56 (d, J = 8.1 Hz, 1H), 7.42 (t, J = 7.6 Hz, 1H), 7.36 (t, J = 7.4 Hz, 1H), 6.73 – 6.64 (m, 2H), 6.51 (dd, J = 7.8, 1.8 Hz, 1H), 6.35 (dd, J = 7.8, 1.8 Hz, 1H), 5.98 (dd, J = 7.6, 1.8 Hz, 1H), 5.54 (dd, J = 7.7, 1.8 Hz, 1H), 4.11 – 4.05 (m, 1H), 3.26 – 3.03 (m, 3H), 2.94 – 2.85 (m, 1H), 2.84 – 2.66 (m, 2H), 2.29 (ddd, J = 13.1, 9.6, 7.3 Hz, 1H) ppm. – ^{13}C NMR (125 MHz, CDCl$_3$) δ = 171.95 (C$_{quat}$), 171.04 (C$_{quat}$), 142.88 (C$_{quat}$), 140.68 (C$_{quat}$), 140.56 (C$_{quat}$), 138.11 (C$_{quat}$), 137.56 (C$_{quat}$), 136.26 (C$_{quat}$), 135.94 (C$_{quat}$), 134.94 (C$_{quat}$), 133.67 (+, C$_{Ar}$H), 132.83 (+, C$_{Ar}$H), 131.76 (+, C$_{Ar}$H), 131.37 (+, C$_{Ar}$H), 130.31 (+, C$_{Ar}$H), 129.16 (+, C$_{Ar}$H), 129.09 (+, C$_{Ar}$H), 128.87 (+, C$_{Ar}$H), 128.80 (+, C$_{Ar}$H), 127.82 (+, C$_{Ar}$H), 127.13 (C$_{quat}$), 126.71 (+, C$_{Ar}$H), 125.88 (C$_{quat}$), 125.78 (+, C$_{Ar}$H), 125.35 (+, C$_{Ar}$H), 124.59 (C$_{quat}$), 122.69 (+, C$_{Ar}$H), 121.03 (+, C$_{Ar}$H), 109.91 (+, C$_{Ar}$H), 35.14 (–, CH$_2$), 33.96 (–, CH$_2$), 33.57 (–, CH$_2$), 33.26 (–, CH$_2$) ppm. – IR (ATR) $\tilde{v}$ = 3030 (vw), 2927 (vw), 2852 (vw), 1587 (vw), 1510 (w), 1444 (w), 1395 (w), 1364 (w), 1173 (w), 1025 (w), 890 (vw), 831 (w), 765 (w), 739 (w), 686 (w), 639 (w), 597 (vw), 527 (vw), 511 (w), 482 (vw), 415 (vw) cm^{-1}. – EA (C$_{43}$H$_{32}$N$_4$) calc. C: 85.40, H: 5.33, N: 9.26; found C: 84.75, H: 5.24, N: 9.34. – HRMS (APCI, C$_{43}$H$_{32}$N$_4$) calc. 605.2700 [M+H]$^+$; found 605.2678 [M+H]$^+$.

9-(4-Nitrophenyl)-9*N*-carbazole (127)

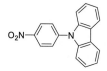

In a 250 mL rund-bottom flask equipped with a reflux condenser carbazole (10.0 g, 60.0 mmol, 1.00 equiv.), 1-fluoro-4-nitrobenzene (12.7 mL, 16.9 g, 120 mmol, 2.00 equiv.) and tripotassium phosphate (28.0 g, 132 mmol, 2.20 equiv.) were dissolved in 120 mL of DMSO and the yellow mixture was stirred at 110 °C for 16 h. After cooling to room temperature, the mixture was diluted with ethyl acetate (300 mL) and washed with brine (3 × 200 mL) to remove DMSO. The organic layer was dried over Na$_2$SO$_4$ and the solvent was removed under reduced pressure. The crude solid was recrystallized from ethanol to yield 12.0 g of the title compound (41.6 mmol, 69%) as a yellow crystalline solid.

R_f = 0.40 (cyclohexane/ethyl acetate; 50:1). – ^{1}H NMR (400 MHz, CDCl$_3$) δ = 8.49 (d, J = 9.0 Hz, 2H), 8.16 (d, J = 7.7 Hz, 2H), 7.81 (d, J = 8.9 Hz, 2H), 7.51 (d, J = 8.2 Hz, 2H), 7.46 (ddd, J = 8.2, 6.9, 1.2 Hz, 2H), 7.36 (ddd, J = 8.1, 7.0, 1.2 Hz, 2H) ppm. – ^{13}C NMR (101 MHz, CDCl$_3$) δ = 145.95 (C$_{quat}$), 143.99 (C$_{quat}$), 139.98 (C$_{quat}$), 126.87 (+, C$_{Ar}$H), 126.62 (+, C$_{Ar}$H), 125.67 (+, C$_{Ar}$H), 124.30 (C$_{quat}$), 121.35 (+, C$_{Ar}$H), 120.77 (+, C$_{Ar}$H), 109.75(+, C$_{Ar}$H) ppm. – IR (ATR) $\tilde{v}$ = 3051 (vw), 1594 (m), 1501 (m), 1451 (m), 1365 (w), 1317 (m), 1228 (m), 1121 (w), 1102 (w), 1003 (w), 915 (w), 853 (w), 840 (w), 741 (m), 718 (m), 690 (m), 638 (w), 617 (w), 564 (w), 527 (w), 480 (w), 422 (w), 410 (w) cm^{-1}. – MS (EI, 70 eV) m/z (%) = 289 (23) [M+H]$^+$, 288 (100) [M]$^+$, 258 (8) [C$_{18}$H$_{14}$N$_2$]$^+$, 243 (13) [C$_{18}$H$_{13}$N]$^+$, 242 (51) [M–NO$_2$]$^+$. – HRMS (EI, C$_{18}$H$_{12}$N$_2$O$_2$) calc. 288.0899; found 288.0898. – EA (C$_{18}$H$_{12}$N$_2$O$_2$) calc. N: 9.72, C: 74.99, H: 4.20; found N: 9.45, C: 74.92, H: 4.26. The analytical data is consistent with literature.[228]

4-Carbazol-9-yl-phenylaniline (128)

To a suspension of 9-(4-nitrophenyl)-9N-carbazole (11.5 g, 40.0 mmol, 1.00 equiv.) in 350 mL of ethanol SnCl$_2 \cdot$2H$_2$O (45.1 g, 200 mmol, 5.00 equiv.) were added quickly. The yellow reaction mixture was heated to reflux for 4 h. The mixture was concentrated under reduced pressure and neutralized by adding NaOH (40 wt%) slowly until the mixture was alkaline.

After extraction with toluene (3 × 100 mL), the combined organic layers were dried over Na$_2$SO$_4$ and the solvent was removed under reduced pressure. The orange crude product was purified by column chromatography (silica, dichloromethane) to yield 8.45 g of the title compound (32.7 mmol, 82%) as a yellow luminescent solid.

R_f = 0.15 (cyclohexane/ethyl acetate; 10:1). – ^{1}H NMR (400 MHz, DMSO-d_6) δ = 8.21 (d, J = 7.7 Hz, 2H), 7.40 (ddd, J = 8.3, 7.0, 1.2 Hz, 2H), 7.28 – 7.21 (m, 4H), 7.18 (d, J = 8.6 Hz, 2H), 6.80 (d, J = 8.6 Hz, 2H), 5.44 (s, 2H) ppm. – ^{13}C NMR (101 MHz, DMSO-d_6) δ = 148.56 (C$_{quat}$), 140.93 (C$_{quat}$), 127.69 (+, C$_{Ar}$H), 125.91 (+, C$_{Ar}$H), 124.43 (C$_{quat}$), 122.17 (C$_{quat}$), 120.31 (+, C$_{Ar}$H), 119.31 (+, C$_{Ar}$H), 114.61 (+, C$_{Ar}$H), 109.57 (+, C$_{Ar}$H) ppm. – IR (ATR) $\tilde{v}$ = 3365 (w, v(NH$_2$)), 3041 (w), 1892 (w), 1620 (m), 1593 (m), 1514 (s), 1477 (m), 1450 (s), 1363 (w), 1335 (m), 1312 (m), 1280 (m), 1229 (s), 1180 (s), 1146 (m), 1117 (m), 1014 (m), 998 (m), 929 (w), 908 (m), 826 (m), 747 (s), 722 (s), 622 (m), 580 (m), 566 (m), 529 (m), 509 (m), 421 (m) cm^{-1}. – HRMS (APCI, C$_{18}$H$_{14}$N$_2$) calc. 259.1230 [M+H]$^+$; found 259.1223 [M+H]$^+$. The analytical data is consistent with literature.[229]

(rac)-N-(4-(Carbazol-9-yl)phenyl)-4-amino[2.2]paracyclophane (129)

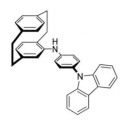

In a 250 mL Schlenk flask, *(rac)*-4-bromo[2.2]paracyclophane (6.89 g, 24.0 mmol, 1.00 equiv.), 4-carbazol-9-yl-phenylaniline (7.75 g, 30.0 mmol, 1.25 equiv.), potassium *tert*-butoxide (3.77 g, 33.6 mmol, 1.40 equiv.), Pd$_2$dba$_3$ (440 mg, 480 µmol, 2 mol%) and SPhos (590 mg, 1.44 mmol, 6 mol%) were evacuated and refilled with argon three times. Then, 150 mL of dry toluene were added, and the mixture was stirred at 120 °C for 16 h. The reaction mixture was diluted with ethyl acetate (150 mL) and washed with saturated NH$_4$Cl solution (3 × 100 mL) and brine (3 × 100 mL). The organic layer was dried over Na$_2$SO$_4$ and the solvent was removed under reduced pressure. The crude solid was purified by recrystallization from isopropanol followed by column chromatography (silica, cyclohexane/ethyl acetate; 10:1) to yield 1.40 g of the title compound (3.01 mmol, 12%) as a yellow powder.

R$_f$ = 0.45 (cyclohexane/ethyl acetate; 10:1). – ^{1}H NMR (400 MHz, DMSO-d_6) δ = 8.23 (d, J = 7.7 Hz, 2H), 8.08 (s, 1H), 7.43 (d, J = 8.3 Hz, 2H), 7.34 (t, J = 9.0 Hz, 4H), 7.26 (d, J = 7.9 Hz, 2H), 7.11 (d, J = 8.8 Hz, 2H), 7.06 (dd, J = 7.7, 1.9 Hz, 1H), 6.58 (dd, J = 7.8, 1.9 Hz, 1H), 6.55 – 6.43 (m, 3H), 6.36 (dd, J = 7.8, 1.8 Hz, 1H), 6.03 (d, J = 1.7 Hz, 1H), 3.15 – 2.82 (m, 7H), 2.78 – 2.69 (m, 1H) ppm. – ^{13}C NMR (101 MHz, DMSO-d_6) δ = 144.06 (C$_{quat}$), 140.66 (C$_{quat}$), 139.96 (C$_{quat}$), 139.12 (C$_{quat}$), 138.52 (C$_{quat}$), 135.67 (+, C$_{Ar}$H), 133.02 (+, C$_{Ar}$H), 132.39 (+, C$_{Ar}$H), 131.90 (C$_{quat}$), 131.70 (+, C$_{Ar}$H), 128.19 (+, C$_{Ar}$H), 127.42 (+, C$_{Ar}$H), 127.19 (+, C$_{Ar}$H), 126.89 (C$_{quat}$), 126.63 (+, C$_{Ar}$H), 125.98 (+, C$_{Ar}$H), 122.28 (C$_{quat}$), 120.34 (+, C$_{Ar}$H), 119.49 (+, C$_{Ar}$H), 115.28 (+, C$_{Ar}$H), 109.59 (+, C$_{Ar}$H), 34.74 (–, CH$_2$), 34.44 (–, CH$_2$), 33.82 (–, CH$_2$), 33.54 (–, CH$_2$) ppm. – IR (ATR) $\tilde{v}$ = 3370 (w, v(NH)), 3032 (vw), 2923 (w), 2849 (w), 1884 (vw), 1594 (w), 1514 (m), 1478 (m), 1450 (m), 1416 (w), 1307 (m), 1231 (m), 1181 (w), 1147 (w), 1106 (w), 998 (vw), 910 (vw), 892 (vw), 827 (w), 795 (w), 746 (m), 725 (m), 653 (w), 624 (w), 567 (vw), 532 (w), 495 (w), 438 (w), 4223 (w) cm^{-1}. – HRMS (APCI, C$_{34}$H$_{28}$N$_2$) calc. 465.2325 [M+H]$^+$; found 465.2309 [M+H]$^+$.

(rac)-[2]Paracyclo[2]6-(N-carbazolyl)(1,4)carbazolophane (130)

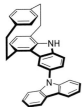

In a 100 mL three-necked flask equipped with a reflux condenser *(rac)*-N-(4-(carbazol-9-yl)phenyl)-4-amino[2.2]paracyclophane (560 mg, 1.20 mmol, 1.00 equiv.), palladium acetate (50.0 mg, 240 μmol, 20 mol%) and pivalic acid (70.0 mg, 720 μmol, 60 mol%) were dissolved in 10 mL of acetic acid and the mixture was stirred at 120 °C for 7 h. Compressed air was added continuously to the suspension using a Teflon hose. The crude product was dissolved in water (50 mL) and the mixture was neutralized with solid NaOH and solid sodium bicarbonate. After extraction with ethyl acetate (3 × 200 mL), the combined organic layers were washed with brine (3 × 100 mL) and dried over Na_2SO_4. The solvent was removed under reduced pressure and the residue was purified by column chromatography (silica, cyclohexane/ethyl acetate; 10:1) to yield 193 mg of the title compound (417 μmol, 34%) as an ochery solid.

R_f = 0.20 (cyclohexane/ethyl acetate; 10:1). – ^{1}H NMR (400 MHz, DMSO-d_6) δ = 11.44 (s, 1H), 8.30 (d, J = 7.7 Hz, 2H), 8.17 (d, J = 2.0 Hz, 1H), 7.78 (d, J = 8.4 Hz, 1H), 7.51 (dd, J = 8.5 Hz, J = 2.0 Hz, 1H), 7.49 – 7.43 (m, 2H), 7.42 – 7.33 (m, 2H), 7.30 (t, J = 7.4 Hz, 2H), 6.67 (d, J = 7.4 Hz, 1H), 6.52 (d, J = 7.4 Hz, 1H), 6.47 (dd, J = 7.9, 1.8 Hz, 1H), 6.36 (dd, J = 7.8, 1.8 Hz, 1H), 5.91 (dd, J = 7.7, 1.8 Hz, 1H, 5.22 (dd, J = 7.7, 1.9 Hz, 1H), 3.85 (dd, J = 9.8 Hz, 1H), 3.62 – 3.52 (m, 1H), 3.12 – 2.89 (m, 5H), 2.86 – 2.75 (m, 1H) ppm. – ^{13}C NMR (101 MHz, DMSO-d_6) δ = 141.52 (C_{quat}), 141.45 (C_{quat}), 138.52 (C_{quat}), 137.40 (C_{quat}), 136.91 (C_{quat}), 135.24 (C_{quat}), 131.93 (+, $C_{Ar}H$), 131.18 (+, $C_{Ar}H$), 127.56 (C_{quat}), 126.16 (+, $C_{Ar}H$), 125.79 (+, $C_{Ar}H$), 125.29 (+, $C_{Ar}H$), 125.15 (C_{quat}), 124.28 (C_{quat}), 124.22 (+, $C_{Ar}H$), 123.77 (+, $C_{Ar}H$), 122.91 (C_{quat}), 122.35 (C_{quat}), 120.79 (+, $C_{Ar}H$), 120.49 (+, $C_{Ar}H$), 119.53 (+, $C_{Ar}H$), 112.22 (+, $C_{Ar}H$), 33.32 (–, CH_2), 33.10 (–, CH_2), 32.80 (–, CH_2), 30.59 (–, CH_2) ppm. – IR (ATR) ṽ = 3415 (vw, v(NH)), 3012 (vw), 2915 (vw), 2848 (vw), 1594 (vw), 1493 (vw), 1473 (w), 1449 (w), 1314 (w), 1232 (w), 992 (vw), 917 (vw), 872 (vw), 806 (w), 750 (w), 724 (w), 651 (w), 605 (vw), 583 (vw), 541 (vw), 520 (w), 492 (vw), 423 (w), 391 (vw) cm^{-1}. – HRMS (APCI, $C_{34}H_{26}N_2$) calc. 463.2169 [M+H]$^+$; found 463.2153 [M+H]$^+$.

(rac)-1-(N-[2]Paracyclo[2]-6-(N-carbazolyl)(1,4)carbazolophanyl)-4-(4,6-diphenyl-1,3,5-triazin-2-yl)-benzene (131)

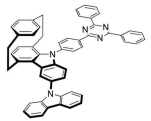

In a sealable vial, (rac)-[2]paracyclo[2]6-(N-carbazolyl)(1,4)carbazolophane (190 mg, 400 µmol, 1.00 equiv.), 2-(4-Fluorophenyl)-4,6-diphenyl-1,3,5-triazine (200 mg, 600 µmol, 1.50 equiv.) and tripotassium phosphate (190 mg, 880 µmol, 2.20 equiv.) were dissolved in 5 mL of DMSO and stirred at 110 °C for 16 h. After cooling to room temperature, the mixture was diluted with ethyl acetate (50 mL) and washed with brine (3 × 30 mL). The organic layer was dried over Na_2SO_4 and the solvent was removed under reduced pressure. The orange residue was purified by column chromatography (silica, gradient cyclohexane/dichloromethane; 50:1 to 10:1 to 5:1 to 2:1) to yield 115 mg of the title compound (149 µmol, 37%) as a yellow, highly luminescent solid.

R_f = 0.34 (cyclohexane/dichloromethane; 10:1). – ^{1}H NMR (400 MHz, CDCl$_3$) δ = 8.85 (dd, J = 8.1 Hz, 1.7 Hz, 4H), 8.29 (d, J = 2.0 Hz, 1H), 8.22 (d, J = 7.8 Hz, 2H), 7.74 (d, J = 8.6 Hz, 1H), 7.69 – 7.58 (m, 8H), 7.56 (dd, J = 8.6, 1.9 Hz, 2H), 7.54 – 7.37 (m, 4H), 7.38 – 7.27 (m, 3H), 6.76 (d, J = 7.6 Hz, 1H), 6.70 (d, J = 7.6 Hz, 1H), 6.53 (dd, J = 7.8, 1.8 Hz, 1H), 6.39 (dd, J = 7.9, 1.3 Hz, 1H), 6.12 (dd, J = 7.8, 1.5 Hz, 1H), 5.76 (dd, J = 7.7, 1.6 Hz, 1H), 3.95 – 3.87 (m, 1H), 3.21 – 2.76 (m, 6H), 2.41 – 2.31 (m, 1H) ppm. – ^{1}H NMR (500 MHz, C$_6$D$_6$) δ = 8.98 – 8.94 (m, 4H), 8.93 – 8.89 (m, 2H), 8.86 (d, J = 8.8 Hz, 1H), 8.27 (d, J = 2.0 Hz, 1H), 8.19 (d, J = 7.8 Hz, 2H), 7.52 (d, J = 8.6 Hz, 1H), 7.45 – 7.28 (m, 14H), 6.63 (d, J = 7.5 Hz, 1H), 6.59 (d, J = 7.5 Hz, 1H), 6.42 (dd, J = 7.8, 1.9 Hz, 1H), 6.29 (dd, J = 7.8, 1.9 Hz, 1H), 6.22 (dd, J = 7.6, 1.9 Hz, 1H), 5.96 (dd, J = 7.6, 1.9 Hz, 1H), 3.56 – 3.48 (m, 1H), 2.97 – 2.89 (m, 2H), 2.87 – 2.78 (m, 2H), 2.72 (ddd, J = 13.7, 9.8, 7.2 Hz, 1H), 2.63 – 2.53 (m, 1H), 2.37 (ddd, J = 13.0, 9.4, 6.7 Hz, 1H) ppm. – ^{13}C NMR (126 MHz, CDCl$_3$) δ = 172.00 (C$_{quat}$), 170.94 (C$_{quat}$), 142.51 (C$_{quat}$), 142.06 (C$_{quat}$), 141.42 (C$_{quat}$), 139.68 (C$_{quat}$), 138.21 (C$_{quat}$), 137.54 (C$_{quat}$), 136.35 (C$_{quat}$), 136.21 (C$_{quat}$), 136.16 (C$_{quat}$), 135.40 (C$_{quat}$), 134.33 (+, C$_{Ar}$H), 132.88 (+, C$_{Ar}$H), 131.98 (+, C$_{Ar}$H), 131.65 (+, C$_{Ar}$H), 130.68 (C$_{quat}$), 130.50 (+, C$_{Ar}$H), 129.17 (+, C$_{Ar}$H), 129.09 (+, C$_{Ar}$H), 128.89 (+, C$_{Ar}$H), 128.80 (+, C$_{Ar}$H), 128.19 (+, C$_{Ar}$H), 126.79 (C$_{quat}$), 126.68 (+, C$_{Ar}$H), 126.14 (+, C$_{Ar}$H), 125.86 (+, C$_{Ar}$H), 125.05 (+, C$_{Ar}$H), 124.88 (C$_{quat}$), 123.25 (C$_{quat}$), 121.85 (+, C$_{Ar}$H), 120.51 (+, C$_{Ar}$H), 119.79 (+, C$_{Ar}$H), 111. (+, C$_{Ar}$H), 109.87 (+, C$_{Ar}$H), 35.16 (–, CH$_2$), 33.88 (–, CH$_2$), 33.48 (–, CH$_2$), 33.30 (–, CH$_2$) ppm. – IR (ATR) $\tilde{v}$ = 3033 (vw), 2921 (vw), 2850 (vw), 1736 (vw), 1587 (w), 1510 (m), 1472 (w), 1444 (m), 1364 (m), 1229 (w), 1171 (w), 1014 (w), 832 (w), 764 (w), 747 (w), 722 (w), 689 (w), 641 (w), 546 (vw), 515 (w) cm^{-1}. – HRMS (APCI, C$_{55}$H$_{39}$N$_5$) calc. 770.3278 [M+H]$^+$; found 770.3256 [M+H]$^+$.

(rac)-*N*-(4-(*tert*-butyl)phenyl)-4-amino[2.2]paracyclophane (133)

 Sealable vials were charged with *(rac)*-4-bromo[2.2]paracyclophane (11.5 g, 40.0 mmol, 1.00 equiv.), 4-*tert*-butylaniline (7.16 g, 48.0 mmol, 1.25 equiv.), potassium *tert*-butoxide (6.28 g, 56.0 mmol, 1.40 equiv.), Pd$_2$dba$_3$ (730 mg, 900 µmol, 2 mol%) and SPhos (990 mg, 2.40 mmol, 6 mol%) in total and evacuated and purged with argon three times. Then, 100 mL of dry toluene were added and the mixture were stirred at 115 °C for 16 h. The combined reaction mixtures were diluted with ethyl acetate (100 mL) and washed with sat. NH$_4$Cl solution (50 mL) and brine (3 × 50 mL). The combined organic layers were dried over Na$_2$SO$_4$ and the solvent was removed under reduced pressure. The crude solid was purified by column chromatography (silica, *n*-hexane:ethyl acetate; 50:1) to yield 6.70 g of the title compound (18.8 mmol, 47%) as a white solid.

R$_f$ = 0.28 (*n*-hexane/ethyl acetate; 50:1). – ^{1}H NMR (500 MHz, CDCl$_3$) δ = 7.26 (d, *J* = 8.6 Hz, 2H), 7.04 (d, *J* = 8.0 Hz, 1H), 6.97 – 6.84 (m, 2H), 6.60 (dd, *J* = 7.8, 2.0 Hz, 1H), 6.49 – 6.41 (m, 3H), 6.35 (d, *J* = 7.7 Hz, 1H), 5.87 (s, 1H), 5.51 (bs, 1H), 3.20 – 2.97 (m, 6H), 2.96 – 2.85 (m, 1H), 2.78 – 2.63 (m, 1H), 1.32 (s, 9H) ppm. – ^{13}C NMR (126 MHz, CDCl$_3$) δ = 143.28 (C$_{quat}$), 141.32 (C$_{quat}$), 141.00 (C$_{quat}$), 140.82 (C$_{quat}$), 139.61 (C$_{quat}$), 139.12 (C$_{quat}$), 135.93 (+, C$_{Ar}$H), 133.54 (+, C$_{Ar}$H), 133.14 (+, C$_{Ar}$H), 132.80 (+, C$_{Ar}$H), 131.45 (+, C$_{Ar}$H), 130.67 (C$_{quat}$), 127.70 (+, C$_{Ar}$H), 126.07 (+, C$_{Ar}$H), 125.17 (+, C$_{Ar}$H), 116.31 (+, C$_{Ar}$H), 35.37 (–, CH$_2$), 35.05 (–, CH$_2$), 34.24 (C$_{quat}$, CMe$_3$), 33.94 (–, CH$_2$), 33.75 (–, CH$_2$), 31.63 (+, CH$_3$) ppm. – IR (ATR) $\tilde{v}$ = 3895 (vw), 3713 (vw), 3638 (vw), 3514 (vw), 3392 (w), 2948 (w), 2451 (vw), 2323 (vw), 2169 (vw), ff1593 (w), 1515 (w), 1302 (w), 1188 (w), 1109 (w), 983 (w), 895 (w), 829 (w), 715 (w), 654 (w), 547 (w), 378 (w), cm^{-1}. – HRMS (ESI, C$_{26}$H$_{29}$N) calc. 356.2373 [M+H]$^+$; found 356.2354 [M+H]$^+$.

(rac)-[2]Paracyclo[2]6-(*tert*-butyl)(1,7)carbazolophane (135)

 In a 250 mL three-necked flask equipped with a reflux condenser *(rac)-N*-(4-(*tert*-butyl)phenyl)-4-amino[2.2]paracyclophane (6.58 g, 18.5 mmol, 1.00 equiv.), 1.04 g of palladium acetate (4.63 mmol, 25 mol%) and pivalic acid (1.42 g, 13.9 mmol, 75 mol%.) were dissolved in 180 mL of acetic acid and the mixture was heated to reflux for 7 h. While stirring, compressed air was added continuously to the suspension using a Teflon hose. The crude product was dissolved in 200 mL of water and 100 mL of ethyl acetate. The mixture was then neutralized by adding solid NaOH and solid sodium bicarbonate and extracted with 400 mL of ethyl acetate three times. The combined organic layers were washed with brine (300 mL) and dried over Na_2SO_4. Then, the solvent was removed under reduced pressure and the black residue was purified by column chromatography (silica, *n*-hexane/ethyl acetate; 5:1) to yield 593 mg of the title compound (1.68 mmol, 9%) as a white solid.

R_f = 0.64 (*n*-hexane/ethyl acetate; 5:1). – ^{1}H NMR (500 MHz, CDCl$_3$) δ = 7.68 (d, J = 1.7 Hz, 1H), 7.67 (d, J = 8.1 Hz, 1H), 7.14 (dd, J = 7.7, 2.1 Hz, 1H), 7.10 (d, J = 1.7 Hz, 1H), 7.01 (dd, J = 7.7, 2.0 Hz, 1H), 6.85 (dd, J = 8.1, 1.3 Hz, 1H), 6.42 (d, J = 1.3 Hz, 1H), 6.39 (bs, 1H), 5.20 (dd, J = 7.7, 2.0 Hz, 1H), 4.46 (dd, J = 7.7, 2.1 Hz, 1H), 3.15 – 2.99 (m, 3H), 2.95 (ddd, J = 13.0, 4.3, 3.0 Hz, 1H), 2.87 – 2.75 (m, 2H), 2.19 – 2.10 (m, 2H), 1.46 (s, 9H) ppm. – ^{13}C NMR (126 MHz, CDCl$_3$) δ = 147.13 (C$_{quat}$), 144.73 (C$_{quat}$), 143.80 (C$_{quat}$), 139.24 (C$_{quat}$), 138.20 (C$_{quat}$), 136.89 (C$_{quat}$), 129.10 (+, C$_{Ar}$H), 128.97 (+, C$_{Ar}$H), 128.95 (+, C$_{Ar}$H), 126.83 (C$_{quat}$), 126.80 (+, C$_{Ar}$H), 125.64 (C$_{quat}$), 123.28 (C$_{quat}$), 122.87 (+, C$_{Ar}$H), 120.53 (+, C$_{Ar}$H), 120.35 (+, C$_{Ar}$H), 119.52 (+, C$_{Ar}$H), 114.10 (+, C$_{Ar}$H), 38.59 (–, CH$_2$), 37.47 (–, CH$_2$), 37.07 (–, CH$_2$), 34.87 (C$_{quat}$), 32.74 (–, CH$_2$), 32.28 (+, CH$_3$) ppm. – IR (ATR) $\tilde{v}$ = 3404 (w), 3030 (vw), 2938 (w), 2849 (vw), 2645 (vw), 2460 (vw), 2346 (vw), 2179 (vw), 2052 (vw), 1911 (vw), 1807 (vw), 1714 (vw), 1591 (vw), 1435 (w), 1360 (w), 1236 (w), 995 (vw), 858 (w), 797 (w), 652 (w), 495 (m) cm^{-1}. – HRMS (ESI, C$_{26}$H$_{27}$N) calc. 354.2216 [M+H]$^+$; found 354.2207 [M+H]$^+$.

(rac)-[2]Paracyclo[2]6-(bromo)(1,4)carbazolophane (150)

In a 100 mL round-bottom flask (rac)-[2]paracyclo[2](1,4)carbazolophane (200 mg, 660 μmol, 1.00 equiv.) was dissolved in a mixture of 5 mL of acetonitrile and 10 mL of dichloromethane and cooled to 0 °C. Under constant stirring, NBS (120 mg, 660 μmol, 1.00 equiv.) was added dropwise *via* syringe as a solution in 5 mL of acetonitrile. The reaction mixture was allowed to warm to room temperature and was stirred for 2 h. The reaction was quenched with water. Dichloromethane (50 mL) was added and the layers were separated. The aqueous phase was extracted with dichloromethane (3 × 30 mL). The combined organic layers were washed with brine (100 mL), dried over Na_2SO_4 and the solvent was removed under reduced pressure. The crude compound was purified by column chromatography (silica, hexanes/ethyl acetace; 15:1) to yield 191 mg of the title compound (508 μmol, 77%) as a white solid.

R_f = 0.19 (hexanes/ethyl acetate; 15:1). – ^{1}H NMR (500 MHz, CDCl$_3$) δ = 8.17 (d, J = 1.8 Hz, 1H), 7.89 (bs, 1H), 7.49 (dd, J = 8.5, 1.8 Hz, 1H), 7.34 (d, J = 8.5 Hz, 1H), 6.65 (d, J = 7.5 Hz, 1H), 6.58 (d, J = 7.5 Hz, 1H), 6.51 (dd, J = 7.8, 1.9 Hz, 1H), 6.37 (dd, J = 7.8, 2.0 Hz, 1H), 5.93 (dd, J = 7.7, 2.0 Hz, 1H), 5.30 (dd, J = 7.7, 2.0 Hz, 1H), 3.99 – 3.89 (m, 1H), 3.39 – 3.26 (m, 1H), 3.20 – 2.90 (m, 6H) ppm. – ^{13}C NMR (126 MHz, CDCl$_3$) δ = 140.58 (C$_{quat}$), 137.99 (C$_{quat}$), 137.47 (C$_{quat}$), 137.43 (C$_{quat}$), 136.07 (C$_{quat}$), 132.28 (+, C$_{Ar}$H), 131.85 (+, C$_{Ar}$H), 131.79 (+, C$_{Ar}$H), 127.74 (+, C$_{Ar}$H), 127.08 (C$_{quat}$), 127.00 (+, C$_{Ar}$H), 126.52 (+, C$_{Ar}$H), 124.92 (+, C$_{Ar}$H), 124.71 (+, C$_{Ar}$H), 124.63 (C$_{quat}$), 122.51 (C$_{quat}$), 112.52 (C$_{quat}$), 112.20 (+, C$_{Ar}$H), 34.01 (–, CH$_2$), 33.72 (–, CH$_2$), 33.33 (–, CH$_2$), 31.17 (–, CH$_2$) ppm.– IR (ATR) $\tilde{v}$ = 3393 (w), 3381 (w), 3031 (vw), 2925 (w), 1591 (w), 1445 (w), 1300 (w), 950 (w), 877 (w), 800 (m), 717 (w), 587 (w), 523 (w), cm^{-1}. – HRMS (ESI, C$_{22}$H$_{18}$BrN) calc. 376.0695 [M(^{79}Br)+H]$^+$; found 376.0685 [M(^{79}Br)+H]$^+$; calc. 378.0675 [M(^{81}Br)+H]$^+$; found 378.0662 [M(^{81}Br)+H]$^+$.

5.4   Unexpected para C–H Activation

(*rac*)-4-Phenylamino-7-phenyl[2.2]paracyclophane (132a)

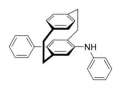

In a sealable vial, (*rac*)-4-phenylamino[2.2]paracyclophane (150 mg, 500 µmol, 1.00 equiv.), potassium *tert*-butoxide (112 mg, 1.00 mmol, 2.00 equiv.) and Pd(PPh₃)₄ (28.9 mg, 5 mol%, 25 µmol) were added. The vial was and evacuated and purged with argon three times. Then, 3 mL of dry toluene and bromobenzene (160 µL, 236 mg, 1.50 mmol, 3.00 equiv.) were added *via* syringe. The reaction mixture was stirred at 120 °C for 16 h. It was quenched with sat. NH₄Cl solution (5 mL) and extracted with ethyl acetate (3 × 10 mL). The combined organic layers were washed with brine (20 mL), dried over Na₂SO₄ and the solvent was removed under reduced pressure. The crude product was purified by column chromatography (silica, *n*-hexane/ethyl acetate; 10:1) to yield 96.0 mg of the title compound (2.57 mmol, 51%) as a white solid.

R_f = 0.41 (*n*-hexane/ethyl acetate; 10:1). – R_f = 0.23 (cyclohexane/toluene; 2:1 + 5 vol% triethylamine). – ¹H NMR (500 MHz, DMSO-d_6) δ = 7.80 (s, 1H), 7.54 – 7.44 (m, 4H), 7.33 (dt, J = 6.0, 3.2 Hz, 1H), 7.18 (t, J = 7.6 Hz, 2H), 7.07 (d, J = 7.7 Hz, 1H), 6.93 (d, J = 8.0 Hz, 2H), 6.76 (t, J = 7.3 Hz, 1H), 6.63 (d, J = 7.9 Hz, 1H), 6.55 – 6.50 (m, 2H), 6.48 (d, J = 7.8 Hz, 1H), 6.03 (s, 1H). 3.28 (ddd, J = 12.9, 9.8, 2.4 Hz, 1H), 3.05 – 2.93 (m, 2H), 2.90 – 2.74 (m, 2H), 2.73 – 2.59 (m, 2H), 2.40 (ddd, J = 12.6, 10.0, 5.9 Hz, 1H) ppm. – ¹³C NMR (126 MHz, DMSO-d_6) δ = 143.78 (C$_{quat}$), 140.91 (C$_{quat}$), 140.11 (C$_{quat}$), 138.98 (C$_{quat}$), 138.40 (C$_{quat}$), 137.15 (C$_{quat}$), 135.67 (C$_{quat}$), 135.02 (+, C$_{Ar}$H), 132.37 (+, C$_{Ar}$H), 131.49 (+, C$_{Ar}$H), 130.11 (C$_{quat}$), 129.54 (+, C$_{Ar}$H), 129.18 (+, C$_{Ar}$H), 128.86 (+, C$_{Ar}$H), 128.56 (+, C$_{Ar}$H), 128.48 (+, C$_{Ar}$H), 127.80 (+, C$_{Ar}$H), 126.31 (+, C$_{Ar}$H), 118.91 (+, C$_{Ar}$H), 115.60 (+, C$_{Ar}$H), 33.98 (–, CH₂), 33.41 (–, CH₂), 33.33 (–, CH₂), 33.09 (–, CH₂) ppm. – IR (ATR) $\tilde{v}$ = 3378 (vw), 2922 (vw), 2849 (vw), 1729 (vw), 1590 (vw), 1557 (vw), 1496 (w), 1480 (vw), 1431 (vw), 1311 (vw), 1249 (vw), 1174 (vw), 1074 (vw), 1030 (vw), 886 (vw), 795 (vw), 769 (vw), 745 (vw), 693 (vw), 641 (vw), 598 (vw), 524 (vw), 486 (vw), 421 (vw) cm⁻¹. – MS (EI, 70 eV) *m/z* (%) = 375 (4) [M]⁺, 270 (7) [C₂₀H₁₇N–H]⁺, 166 (47) [C₁₂H₈N]⁺, 105 (10) [C₈H₉]⁺. – HRMS (EI, C₂₂H₂₁N): calc. 375.1987; found 375.1988.

General Procedure of C–H Activation using liquid bromoarenes (GP1)

Ia sealable vial, (*rac*)-4-phenylamino[2.2]paracyclophane (150 mg, 500 µmol, 1.00 equiv.), potassium *tert*-butoxide (70.1 mg, 625 µmol, 1.25 equiv.) and Pd(PPh₃)₄ (28.9 mg, 5 mol%, 25 µmol) were added. The vial was and evacuated and purged with argon three times. Then, 3 mL of dry toluene and the respective liquid bromoarene (625 µmol, 1.25 equiv.) was added *via* syringe. The reaction mixture was stirred at 120 °C for 16 h. It was quenched with sat. NH₄Cl solution (5 mL) and extracted with ethyl acetate (3 × 10 mL). The combined organic layers were washed with brine (20 mL), dried over Na₂SO₄ and the solvent was removed under reduced pressure. The crude product was purified by column chromatography.

(*rac*)-4-*N*-Phenylamino-7-(2-methoxy)phenyl[2.2]paracyclophane (132l)

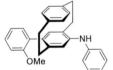

The crude product was purified by column chromatography (silica, *n*-hexane/ethyl acetate; 10:1) to yield 88.3 mg of the title compound (218 µmol, 44%) as a white impure solid.

R_f = 0.26 (*n*-hexane/ethyl acetate; 10:1). – ¹H NMR (500 MHz, C₆D₆) δ = 7.64 (dd, *J* = 7.5, 1.8 Hz, 1H), 7.29 – 7.21 (m, 2H), 7.15 – 7.08 (m, 2H), 7.01 (dd, *J* = 7.8, 2.0 Hz, 1H), 6.87 (dd, *J* = 7.9, 1.9 Hz, 2H), 6.83 (dd, *J* = 7.6, 2.2 Hz, 2H), 6.69 (dd, *J* = 8.2, 1.1 Hz, 1H), 6.53 (s, 1H), 6.40 (dd, *J* = 7.9, 1.8 Hz, 1H), 6.26 (dd, *J* = 7.8, 1.9 Hz, 1H), 5.63 (s, 1H), 5.13 (s, 1H), 3.30 (s, 3H, OMe), 3.09 (ddd, *J* = 13.4, 9.1, 4.7 Hz, 1H), 3.02 – 2.77 (m, 3H), 2.71 (ddd, *J* = 13.2, 9.9, 5.1 Hz, 1H), 2.65 – 2.59 (m, 2H), 2.39 (ddd, *J* = 13.6, 10.1, 7.0 Hz, 1H) ppm. – MS (APCI) *m/z* (%) = 406 [M+H]⁺.

(*rac*)-4-*N*-Phenylamino-7-(2,5-dimethyl)phenyl[2.2]paracyclophane (132j)

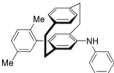

The crude product was purified by column chromatography (silica, *n*-hexane/ethyl acetate; 20:1) to yield 166 mg of the title compound (411 μmol, 82%) as a white solid.

R_f = 0.36 (*n*-hexane/ethyl acetate; 20:1). – ^{1}H NMR (500 MHz, C$_6$D$_6$) δ = 7.71 (s, 1H), 7.25 – 7.20 (m, 3H), 7.17 (dd, *J* = 7.7, 1.8 Hz, 1H), 7.13 (dd, *J* = 7.8, 1.9 Hz, 1H), 7.05 – 6.98 (m, 3H), 6.94 (t, *J* = 7.3 Hz, 1H), 6.59 (s, 1H), 6.55 (dd, *J* = 7.9, 1.8 Hz, 1H), 6.34 (dd, *J* = 7.8, 1.9 Hz, 1H), 5.63 (s, 1H), 5.23 (s, 1H), 3.05 (ddd, *J* = 13.8, 9.4, 1.9 Hz, 1H), 2.99 – 2.84 (m, 3H), 2.78 – 2.57 (m, 3H), 2.54 (s, 3H, Me), 2.48 (ddd, *J* = 13.9, 10.2, 7.0 Hz, 1H), 2.30 (s, 3H, Me) ppm. – ^{13}C NMR (126 MHz, C$_6$D$_6$) δ = 143.81 (C$_{quat}$), 141.72 (C$_{quat}$), 139.84 (C$_{quat}$), 139.53 (C$_{quat}$), 139.46 (C$_{quat}$), 139.19 (C$_{quat}$), 136.04 (+, C$_{Ar}$H), 135.91 (C$_{quat}$), 135.45 (C$_{quat}$), 133.83 (C$_{quat}$), 133.33 (+, C$_{Ar}$H), 131.39 (+, C$_{Ar}$H), 130.83 (C$_{quat}$), 130.59 (+, C$_{Ar}$H), 130.13 (+, C$_{Ar}$H), 129.52 (+, C$_{Ar}$H), 128.35 (+, C$_{Ar}$H), 128.16 (+, C$_{Ar}$H), 127.97 (+, C$_{Ar}$H), 126.90 (+, C$_{Ar}$H), 120.63 (+, C$_{Ar}$H), 116.76 (+, C$_{Ar}$H), 34.94 (–, CH$_2$), 33.75 (–, CH$_2$), 33.73 (–, CH$_2$), 33.10 (–, CH$_2$), 21.48 (+, CH$_3$), 20.23 (+, CH$_3$) ppm. – IR (ATR) $\tilde{v}$ = 3988 (w), 3836 (w), 3607 (w), 3372 (w), 2921 (w, NH), 2849 (w), 2322 (w), 2171 (vw), 1741 (w), 1588 (w), 1495 (m), 1308 (w), 1247 (w), 1076 (w), 881 (w), 820 (m), 750 (m), 605 (w), 461 (m) cm^{-1}. – MS (FAB, 3-NBA) *m/z* = 403 [M]$^+$. – HRMS (FAB, 3-NBA, C$_{30}$H$_{29}$N): calc. 403.2300; found 403.2299.

(*rac*)-4-*N*-Phenylamino-7-(4-methoxy)phenyl[2.2]paracyclophane (132k)

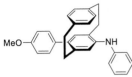

The crude product was purified by column chromatography (silica, *n*-hexane/ethyl acetate; 20:1) to yield 82.0 mg of the title compound (202 μmol, 39%) as a white solid.

R_f = 0.19 (*n*-hexane/ethyl acetate; 20:1). – ^{1}H NMR (500 MHz, C$_6$D$_6$) δ = 7.45 (d, J = 8.6 Hz, 2H), 7.16 – 7.13 (m, 2H), 7.02 – 6.96 (m, 3H), 6.94 – 6.88 (m, 2H), 6.85 (t, J = 7.3 Hz, 1H), 6.81 (dd, J = 7.9, 1.9 Hz, 1H), 6.50 (s, 1H), 6.40 (dd, J = 7.8, 1.8 Hz, 1H), 6.26 (dd, J = 7.8, 1.9 Hz, 1H), 5.68 (s, 1H), 5.14 (s, 1H), 3.43 (s, 3H), 3.42 – 3.38 (m, 1H), 2.97 (ddd, J = 13.7, 9.2, 2.0 Hz, 1H), 2.91 – 2.78 (m, 2H), 2.67 – 2.47 (m, 3H), 2.39 (ddd, J = 13.7, 10.1, 7.2 Hz, 1H) ppm. – ^{13}C NMR (126 MHz, C$_6$D$_6$) δ = 159.20 (C$_{quat}$), 144.15 (C$_{quat}$), 139.54 (C$_{quat}$), 139.41 (C$_{quat}$), 139.38 (C$_{quat}$), 138.08 (C$_{quat}$), 137.42 (C$_{quat}$), 135.61 (+, C$_{Ar}$H), 134.41 (C$_{quat}$), 133.00 (+, C$_{Ar}$H), 131.61 (+, C$_{Ar}$H), 131.27 (C$_{quat}$), 130.92 (+, C$_{Ar}$H), 130.42 (+, C$_{Ar}$H), 129.55 (+, C$_{Ar}$H), 128.88 (+, C$_{Ar}$H), 128.35 (+, C$_{Ar}$H), 128.16 (+, C$_{Ar}$H), 120.57 (+, C$_{Ar}$H), 116.63 (+, C$_{Ar}$H), 114.45 (+, C$_{Ar}$H), 54.90 (+, OCH$_3$), 34.65 (–, CH$_2$), 34.04 (–, CH$_2$), 33.89 (–, CH$_2$), 33.78 (–, CH$_2$) ppm. – IR (ATR) $\tilde{v}$ = 3361 (vw), 2930 (w, NH), 2635 (vw), 2534 (vw), 2291 (vw), 2167 (vw), 2086 (vw), 1901 (vw), 1781 (vw), 1605 (w), 1484 (w), 1241 (w), 1174 (w), 1035 (w), 836 (w), 750 (w), 620 (w), 487 (w) cm^{-1}. – MS (FAB, 3-NBA) m/z = 405 [M]$^+$. – HRMS (FAB, 3-NBA, C$_{29}$H$_{27}$N): calc. 405.2093; found 405.2093.

General Procedure of C–H Activation protocol using solid bromoarenes (GP2)

Ia sealable vial, (*rac*)-4-phenylamino[2.2]paracyclophane (150 mg, 500 µmol, 1.00 equiv.), potassium *tert*-butoxide (70.1 mg, 625 µmol, 1.25 equiv.), Pd(PPh$_3$)$_4$ (28.9 mg, 5 mol%, 25 µmol) and the respective solid bromoarene (625 µmol, 1.25 equiv.) were added. The vial was evacuated and purged with argon three times. Then, 3 mL of dry toluene was added *via* syringe. The reaction mixture was stirred at 120 °C for 16 h. It was quenched with sat. NH$_4$Cl solution (5 mL) and extracted with ethyl acetate (3 × 10 mL). The combined organic layers were washed with brine (20 mL), dried over Na$_2$SO$_4$ and the solvent was removed under reduced pressure. The crude product was purified by column chromatography.

(*rac*)-4-*N*-Phenylamino-7-(4-*N*,*N*-diphenylamino)phenyl[2.2]paracyclophane (132h)

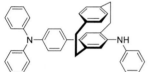

The crude product was purified by column chromatography (silica, *n*-hexane/dichloromethane; 5:1) to yield 32.6 mg of the title compound (284 µmol, 12%) as a white solid.

R$_f$ = 0.29   (*n*-hexane/dichloromethane;   5:1).   – ^{1}H NMR (400 MHz, DMSO-d_6) δ = 7.76 (s, 1H), 7.40 (d, J = 8.6 Hz, 2H), 7.37 – 7.32 (m, 4H), 7.17 (dd, J = 8.5, 7.2 Hz, 2H), 7.12 – 7.02 (m, 9H), 6.97 – 6.88 (m, 2H), 6.79 – 6.72 (m, 1H), 6.62 – 6.53 (m, 2H), 6.52 – 6.45 (m, 2H), 6.02 (s, 1H), 3.06 – 2.91 (m, 2H), 2.90 – 2.74 (m, 2H), 2.71 – 2.57 (m, 2H), 2.50 – 2.41 (m, 2H) ppm. – ^{13}C NMR (101 MHz, DMSO-d_6) δ = 147.24 (C$_{quat}$), 145.42 (C$_{quat}$), 143.85 (C$_{quat}$), 139.83 (C$_{quat}$), 138.91 (C$_{quat}$), 138.41 (C$_{quat}$), 137.07 (C$_{quat}$), 135.30 (C$_{quat}$), 135.24 (C$_{quat}$), 134.76 (+, C$_{Ar}$H), 132.23 (+, C$_{Ar}$H), 131.49 (+, C$_{Ar}$H), 130.11 (C$_{quat}$), 130.04 (+, C$_{Ar}$H), 129.61 (+, C$_{Ar}$H), 129.57 (+, C$_{Ar}$H), 128.81 (+, C$_{Ar}$H), 128.49 (+, C$_{Ar}$H), 127.86 (+, C$_{Ar}$H), 123.98 (+, C$_{Ar}$H), 123.16 (+, C$_{Ar}$H), 123.03 (+, C$_{Ar}$H), 118.78 (+, C$_{Ar}$H), 115.47 (+, C$_{Ar}$H), 33.98 (–, CH$_2$), 33.62 (–, CH$_2$), 33.38 (–, CH$_2$), 33.06 (–, CH$_2$) ppm. – MS (APCI) m/z = 543 [M+H]$^+$.

(*rac*)-4-*N*-Phenylamino-7-(2-anthracenyl)phenyl[2.2]paracyclophane (132g)

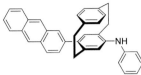

The crude product was purified by column chromatography (silica, *n*-hexane/ethyl acetate; 20:1) to yield 134 mg of the title compound (284 µmol, 57%) as a white solid.

R_f = 0.28 (*n*-hexane/ethyl acetate; 20:1). – ^{1}H NMR (400 MHz, DMSO-d_6) δ = 8.71 (s, 1H), 8.61 (s, 1H), 8.21 – 8.15 (m, 2H), 8.14 – 8.09 (m, 2H), 7.84 (s, 1H), 7.67 (dd, J = 8.7, 1.7 Hz, 1H), 7.56 – 7.49 (m, 2H), 7.20 (dd, J = 8.5, 7.2 Hz, 2H), 7.10 (dd, J = 7.8, 1.8 Hz, 1H), 7.01 – 6.93 (m, 2H), 6.78 (m, 2H), 6.72 (s, 1H), 6.65 – 6.60 (m, 1H), 6.51 (dd, J = 7.8, 1.8 Hz, 1H), 6.11 (s, 1H), 3.48 – 3.37 (m, 1H), 3.11 – 3.00 (m, 2H), 2.91 (ddd, J = 12.8, 9.3, 6.9 Hz, 1H), 2.85 – 2.69 (m, 3H), 2.44 – 2.37 (m, 1H) ppm. – ^{13}C NMR (101 MHz, DMSO-d_6) δ = 143.69 (C_{quat}), 140.46 (C_{quat}), 139.01 (C_{quat}), 138.40 (C_{quat}), 137.93 (C_{quat}), 137.58 (C_{quat}), 135.27 (C_{quat}), 135.24 (+, $C_{Ar}H$), 132.41 (+, $C_{Ar}H$), 131.71 (C_{quat}), 131.50 (+, $C_{Ar}H$), 131.45 (C_{quat}), 131.13 (C_{quat}), 130.10 (C_{quat}), 129.99 (C_{quat}), 129.62 (+, $C_{Ar}H$), 128.83 (+, $C_{Ar}H$), 128.47 (+, $C_{Ar}H$), 128.38 (+, $C_{Ar}H$), 128.10 (+, $C_{Ar}H$), 128.01 (+, $C_{Ar}H$), 127.95 (+, $C_{Ar}H$), 127.73 (+, $C_{Ar}H$), 126.81 (+, $C_{Ar}H$), 126.15 (+, $C_{Ar}H$), 125.69 (+, $C_{Ar}H$), 125.59 (+, $C_{Ar}H$), 125.36 (+, $C_{Ar}H$), 119.03 (+, $C_{Ar}H$), 115.81 (+, $C_{Ar}H$), 34.00 (–, CH_2), 33.61 (–, CH_2), 33.33 (–, CH_2), 33.13 (–, CH_2) ppm. – IR (ATR) $\tilde{v}$ = 3051 (vw), 2921 (w, NH), 2851 (w), 1621 (w), 1449 (w), 1271 (w), 1147 (w), 956 (w), 882 (m), 724 (m), 602 (w), 473 (m) cm^{-1}. – MS (FAB, 3-NBA) m/z = 475 [M]$^+$. – HRMS (FAB, 3-NBA, $C_{36}H_{29}N$): calc. 475.2300; found 475.2300.

(*rac*)-4-*N*-Phenylamino-7-(4-*N*-carbazolyl)phenyl[2.2]paracyclophane (132i)

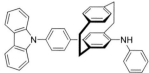

The crude product was purified by column chromatography (silica, *n*-hexane/ethyl acetate; 20:1) to yield 32.6 mg of the title compound (284 µmol, 12%) as a white solid.

R_f = 0.29 (*n*-hexane/ethyl acetate; 20:1). – ^{1}H NMR (500 MHz, C$_6$D$_6$) δ = 8.13 (dd, J = 7.6, 1.0 Hz, 2H), 7.57 – 7.49 (m, 4H), 7.42 – 7.34 (m, 4H), 7.32 – 7.27 (m, 2H), 7.19 – 7.16 (m, 2H), 7.02 (dd, J = 7.8, 1.9 Hz, 1H), 6.97 – 6.92 (m, 2H), 6.88 (t, J = 7.3 Hz, 1H), 6.78 (dd, J = 7.9, 1.9 Hz, 1H), 6.53 (s, 1H), 6.39 (dd, J = 7.9, 1.8 Hz, 1H), 6.28 (dd, J = 7.9, 1.9 Hz, 1H), 5.72 (s, 1H), 5.18 (s, 1H), 3.38 (ddd, J = 12.7, 10.2, 3.0 Hz, 1H), 3.00 – 2.87 (m, 2H), 2.82 (ddd, J = 13.1, 9.3, 7.1 Hz, 1H), 2.70 – 2.61 (m, 1H), 2.61 – 2.48 (m, 2H), 2.43 (ddd, J = 13.8, 10.1, 7.1 Hz, 1H) ppm. – ^{13}C NMR (126 MHz, C$_6$D$_6$) δ = 143.64 (C$_{quat}$), 141.55 (C$_{quat}$), 140.86 (C$_{quat}$), 140.50 (C$_{quat}$), 139.32 (C$_{quat}$), 139.28 (C$_{quat}$), 138.30 (C$_{quat}$), 136.51 (C$_{quat}$), 136.20 (C$_{quat}$), 135.76 (+, C$_{Ar}$H), 133.02 (+, C$_{Ar}$H), 131.66 (+, C$_{Ar}$H), 131.04 (+, C$_{Ar}$H), 130.53 (C$_{quat}$), 130.41 (+, C$_{Ar}$H), 129.61 (+, C$_{Ar}$H), 128.35 (+, C$_{Ar}$H), 128.26 (+, C$_{Ar}$H), 127.45 (+, C$_{Ar}$H), 126.40 (+, C$_{Ar}$H), 124.15 (C$_{quat}$), 121.09 (+, C$_{Ar}$H), 120.84 (+, C$_{Ar}$H), 120.52 (+, C$_{Ar}$H), 117.26 (+, C$_{Ar}$H), 110.33 (+, C$_{Ar}$H), 34.80 (–, CH$_2$), 33.97 (–, CH$_2$), 33.81 (–, CH$_2$), 33.67 (–, CH$_2$) ppm.– IR (ATR) $\tilde{v}$ = 3990 (w), 3891 (w), 3729 (w), 3532 (w), 3378 (w), 3035 (w), 2922 (m, NH), 2325 (w), 1727 (w), 1590 (m), 1449 (m), 1313 (m), 1228 (m), 1017 (w), 906 (w), 837 (w), 746 (s), 498 (m), 422 (m) cm^{-1}. – MS (FAB, 3-NBA) *m/z* = 540 [M]$^+$. – HRMS (FAB, 3-NBA, C$_{40}$H$_{32}$N$_2$): calc. 540.2565; found 540.2566.

(*rac*)-4-*N*,*N*-Diphenylamino-7-phenyl[2.2]paracyclophane (133)

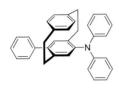

In a sealable vial, (*rac*)-4-phenylamino[2.2]paracyclophane (150 mg, 500 μmol, 1.00 equiv.), potassium *tert*-butoxide (112 mg, 1.00 mmol, 2.00 equiv.) and Pd(PPh$_3$)$_4$ (28.9 mg, 5 mol%, 25 μmol) were added. The vial was and evacuated and purged with argon three times. Then, 3 mL of dry toluene and bromobenzene (160 μL, 236 mg, 1.50 mmol, 3.00 equiv.)

were added *via* syringe. The reaction mixture was stirred at 120 °C for 16 h. It was quenched with sat. NH$_4$Cl solution (5 mL) and extracted with ethyl acetate (3 × 10 mL). The combined organic layers were washed with brine (20 mL), dried over Na$_2$SO$_4$ and the solvent was removed under reduced pressure. The crude product was purified by column chromatography (silica, *n*-hexane/ethyl acetate; 10:1) to yield 20.3 mg of the title compound (45.0 μmol, 9%) as a white impure solid.

R$_f$ = 0.52 (*n*-hexane/ethyl acetate; 10:1). – ^{1}H NMR (500 MHz, CDCl$_3$) δ = 7.53 (d, *J* = 8.3 Hz, 2H), 7.50 – 7.44 (m, 2H), 7.38 – 7.29 (m, 5H), 7.25 (s, 1H), 7.21 – 7.07 (m, 5H), 6.79 (d, *J* = 7.8 Hz, 1H), 6.76 (d, *J* = 7.9 Hz, 1H), 6.47 (s, 1H), 6.42 (d, *J* = 7.8 Hz, 1H), 6.33 (t, *J* = 7.2 Hz, 1H), 5.83 (s, 1H), 3.17 (ddd, *J* = 12.7, 9.5, 5.2 Hz, 1H), 3.10 – 2.81 (m, 4H), 2.80 – 2.69 (m, 1H), 2.67 – 2.45 (m, 2H) ppm. – IR (ATR) $\tilde{v}$ = 2922 (vw), 1584 (w), 1478 (w), 1386 (vw), 1241 (w), 1172 (vw), 1072 (vw), 1029 (vw), 905 (vw), 845 (vw), 795 (vw), 757 (w), 717 (vw), 695 (w), 619 (vw), 587 (vw), 514 (w), 496 (w) cm^{-1}. – MS (APCI) *m/z* (%) = 452 [M+H]$^+$.

5.5   Analytical Data of Chemical Vapor Deposition Project

(*rac*)-4-Acetyl[2.2]paracyclophane (137)

To a mixture of aluminium chloride (18.7 g, 140 mmol, 1.75 equiv.) in 500 mL of dry dichloromethane, 11.4 mL of acetyl chloride (12.6 g, 160 mmol, 2.00 equiv.) was added. After cooling to –78 °C, a solution of [2.2]paracyclophane (16.7 g, 80.0 mmol, 1.00 equiv.) in 250 mL of dry dichloromethane was added dropwise. The reaction mixture was warmed to –18 °C and stirred for another 3 h. The reaction mixture was filtered through glass wool, poured on ice and stirred for 30 min. The layers were separated and the organic layer was washed with brine (400 mL). The solvent was removed under reduced pressure and the crude solid was purified by column chromatography (silica, cyclohexane/ethyl acetate; 40:1) to yield 10.0 g of the title compound (43.9 mmol, 55%) as a white solid. R_f = 0.24 (cyclohexane/ethyl acetate; 40.1). – ^{1}H NMR (400 MHz, CDCl$_3$) δ = 6.93 (d, J = 1.8 Hz, 1H), 6.66 (dd, J = 7.8, 1.8 Hz, 1H), 6.59 – 6.50 (m, 3H), 6.48 (dd, J = 7.9, 1.9 Hz, 1H), 6.39 (dd, J = 7.9, 1.9 Hz, 1H), 3.98 (ddd, J = 12.1, 7.0, 4.5 Hz, 1H), 3.26 – 3.12 (m, 4H), 3.10 – 2.96 (m, 2H), 2.90 – 2.80 (m, 1H), 2.47 (s, 3H) ppm. – ^{13}C NMR (101 MHz, CDCl$_3$) δ = 200.33 (C$_{quat}$, CO), 141.62 (C$_{quat}$), 140.36 (C$_{quat}$), 139.80 (C$_{quat}$), 139.24 (C$_{quat}$), 137.92 (C$_{quat}$), 136.48 (+, C$_{Ar}$H), 136.42 (+, C$_{Ar}$H), 134.29 (+, C$_{Ar}$H), 133.11 (+, C$_{Ar}$H), 132.95 (+, C$_{Ar}$H), 132.09 (+, C$_{Ar}$H), 131.22 (+, C$_{Ar}$H), 36.14 (–, CH$_2$), 35.26 (–, CH$_2$), 35.22 (–, CH$_2$), 34.98 (–, CH$_2$), 28.83 (+, CH$_3$) ppm. – IR (ATR) ṽ = 3343 (vw), 2921 (w), 2849 (w), 2322 (vw), 2230 (vw), 2165 (vw), 2053 (vw), 1897 (vw), 1679 (m), 1552 (w), 1349 (w), 1265 (m), 1176 (w), 1065 (w), 853 (w), 730 (w), 615 (m), 510 (m), 396,66 (w) cm^{-1}.– HRMS (ESI, C$_{18}$H$_{18}$O) calc. 251.1430 [M+H]$^+$, found 251.1418 [M+H]$^+$. The analytical data is consistent with literature.[142]

(*rac*)-4-Formyl[2.2]paracyclophane (5)

To a solution of [2.2]paracyclophane (14.6 g, 70.0 mmol, 1.00 equiv.) in 600 mL of dichloromethane at 0 °C was added 15.4 mL of titanium tetrachloride (26.6 g, 140 mmol, 2.00 equiv.) followed 6.42 mL of 1,1-dichloromethylether (8.16 g, 71.0 mmol, 1.05 equiv.). After stirring for 15 min at 0 °C, the mixture was stirred for 16 h at room temperature. The mixture changes color from clear to yellow to black. The black reaction mixture was poured on ice and stirred for 2 h again changing its color to yellow. The organic layer was separated and the aqueous layer was extracted with dichloromethane (3 × 200 mL). The combined organic layers were washed with brine (300 mL), dried over Na_2SO_4 and the solvent was removed under reduced pressure. The crude compound was filtered (silica, dichloromethane) to remove residual titanium salts and recrystallized from *n*-hexane to yield 14.8 g of the title compound (62.8 mmol, 90%) as a white solid.

R_f = 0.41 (cyclohexane/ethyl acetate; 20.1). – ^{1}H NMR (400 MHz, CDCl$_3$) δ = 9.95 (s, 1H, CHO), 7.02 (d, J = 1.9 Hz, 1H), 6.73 (dd, J = 7.8, 1.9 Hz, 1H), 6.61 – 6.55 (m, 2H), 6.50 (dd, J = 7.8, 1.9 Hz, 1H), 6.43 (dd, J = 7.8, 1.8 Hz, 1H), 6.38 (dd, J = 7.8, 1.8 Hz, 1H), 4.11 (ddd, J = 13.0, 9.9, 1.7 Hz, 1H), 3.35 – 3.14 (m, 3H), 3.14 – 2.89 (m, 4H) ppm. – ^{13}C NMR (101 MHz, CDCl$_3$) δ = 192.02 (+, CHO), 143.30 (C$_{quat}$), 140.73 (C$_{quat}$), 139.57 (C$_{quat}$), 139.52 (C$_{quat}$), 138.16 (+, C$_{Ar}$H), 136.64 (C$_{quat}$), 136.43 (+, C$_{Ar}$H), 136.20 (+, C$_{Ar}$H), 133.34 (+, C$_{Ar}$H), 133.00 (+, C$_{Ar}$H), 132.44 (+, C$_{Ar}$H), 132.23 (+, C$_{Ar}$H), 35.34 (–, CH$_2$), 35.23 (–, CH$_2$), 35.06 (–, CH$_2$), 33.71 (–, CH$_2$) ppm. – IR (ATR) $\tilde{v}$ = 3337 (vw), 3035 (vw), 2922 (w), 2850 (vw), 2759 (vw), 2320 (vw), 2168 (vw), 2041 (vw), 1894 (vw), 1676 (m), 1589 (w), 1489 (w), 1411 (w), 1227 (w), 1145 (w), 875 (w), 797 (w), 719 (w), 636 (w), 517 (w), cm^{-1}. – HRMS (ESI, C$_{17}$H$_{16}$O) calc 237.1274 [M+H]$^+$; found 237.1262 [M+H]$^+$. The analytical data is consistent with literature.[48]

(*Sp,R*)-4-(*N*-1-(Phenylethyl)methaniminyl)-[2.2]paracyclophane (48)

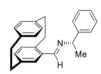

A solution of (*rac*)-4-formyl[2.2]paracyclophane (1.72 g, 7.30 mmol, 1.00 equiv.) and 940 μL of (*R*)-phenylethylamine (880 mg, 7.30 mmol, 1.00 equiv.) in 60 mL of toluene was heated to reflux for 16 h. After cooling to r.t., the solvent was removed under reduced pressure and the residue was repeatedly recrystallized from *n*-hexane until clear cubic crystals formed. The title compound (323 mg, 952 μmol, 26% based on 0.50 equiv. of (*Sp*)-starting material) was obtained as clear crystals.

^{1}H NMR (400 MHz, CDCl$_3$) δ = 8.38 (s, 1H), 7.53 (d, *J* = 7.7 Hz, 2H), 7.41 (t, *J* = 7.7 Hz, 2H), 7.32 – 7.25 (m, 1H), 7.01 (d, *J* = 1.9 Hz, 1H), 6.59 (dd, *J* = 7.7, 2.0 Hz, 1H), 6.54 – 6.47 (m, 3H), 6.40 (d, *J* = 8.0 Hz, 1H), 6.30 (d, *J* = 8.0 Hz, 1H), 4.58 (q, *J* = 6.6 Hz, 1H), 3.99 – 3.79 (m, 1H), 3.24 – 2.97 (m, 6H), 2.87 (ddd, *J* = 9.3, 6.9, 5.2 Hz, 2H), 1.67 (d, *J* = 6.6 Hz, 3H) ppm. – IR (ATR) ṽ = 2919 (vw), 2846 (vw), 2635 (vw), 2503 (vw), 2292 (vw), 2054 (vw), 1895 (vw), 1795 (vw), 1638 (w), 1448 (vw), 1274 (vw), 1068 (vw), 906 (vw), 696 (w), 483 (w), 402 (vw), cm^{-1}. – HRMS (ESI, C$_{25}$H$_{25}$N) calc. 340.2060 [M+H]$^+$, found 340.2042 [M+H]$^+$. The analytical data is consistent with literature.[48]

(*Sp*)-4-Formyl[2.2]paracyclophane (5)

To a solution of (*Sp,R*)-4-(*N*-1-(phenylethyl)methaniminyl)-[2.2]paracyclophane (225 mg, 952 μmol, 1.00 equiv.) in dichloromethane, 5 g of silica was added and the suspension was stirred at room temperature for 30 min. The solvent was removed in vacuo and the resulting solid was purified by short column chromatography (silica, dichloromethane) to yield 225 mg, of the title compound (952 μmol, quant.) as a white solid.

The analytical data is consistent with literature, see above.[48]

(S$_P$,S)/(S$_P$,R)-1-(4-[2.2]Paracyclophanyl)ethanol (139, 141)

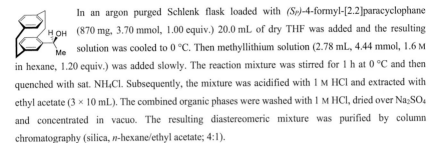

In an argon purged Schlenk flask loaded with *(S$_P$)*-4-formyl-[2.2]paracyclophane (870 mg, 3.70 mmol, 1.00 equiv.) 20.0 mL of dry THF was added and the resulting solution was cooled to 0 °C. Then methyllithium solution (2.78 mL, 4.44 mmol, 1.6 M in hexane, 1.20 equiv.) was added slowly. The reaction mixture was stirred for 1 h at 0 °C and then quenched with sat. NH₄Cl. Subsequently, the mixture was acidified with 1 M HCl and extracted with ethyl acetate (3 × 10 mL). The combined organic phases were washed with 1 M HCl, dried over Na₂SO₄ and concentrated in vacuo. The resulting diastereomeric mixture was purified by column chromatography (silica, *n*-hexane/ethyl acetate; 4:1).

The *(S$_P$,S)* title product **139** was obtained as a white solid in 47% yield (437 mg, 1.73 mmol).
R_f = 0.34 (*n*-hexane/ethyl acetate; 4:1). – ¹H NMR (500 MHz, DMSO-d_6) δ = 6.66 (dd, J = 7.7, 1.8 Hz, 1H), 6.59 (s, 1H), 6.46 (qd, J = 7.8, 1.8 Hz, 2H), 6.39 (dd, J = 7.7, 1.8 Hz, 1H), 6.35 (dd, J = 7.6, 1.9 Hz, 1H), 6.31 (d, J = 7.6 Hz, 1H), 4.89 (d, J = 5.0 Hz, 1H, O*H*), 4.76 (p, J = 6.3 Hz, 1H, C*H*OH), 3.23 (ddd, J = 13.3, 8.7, 3.5 Hz, 1H), 3.06 – 2.87 (m, 6H), 2.75 (ddd, J = 13.4, 9.9, 6.7 Hz, 1H), 1.12 (d, J = 6.4 Hz, 3H, C*H₃*). – ¹³C NMR (126 MHz, DMSO-d_6) δ = 146.03 (C$_{quat}$), 139.22 (C$_{quat}$), 138.89 (C$_{quat}$), 138.87 (C$_{quat}$), 134.59 (+, C$_{Ar}$H), 134.43 (C$_{quat}$), 132.89 (+, C$_{Ar}$H), 132.82 (+, C$_{Ar}$H), 132.42 (+, C$_{Ar}$H), 130.68 (+, C$_{Ar}$H), 130.44 (+, C$_{Ar}$H), 128.78 (+, C$_{Ar}$H), 65.85 (+, CH), 34.83 (–, CH₂), 34.82 (–, CH₂), 33.78 (–, CH₂), 32.69 (–, CH₂), 26.38 (+, CH₃). – IR (ATR) ṽ = 3614 (w), 3352 (w, br), 2966 (w), 2923 (w), 2847 (w), 1591 (w), 1498 (vw), 1438 (w), 1413 (w), 1371 (w), 1319 (vw), 1272 (vw), 1181 (vw), 1144 (w), 1120 (w), 1059 (w), 1023 (w), 933 (vw), 905 (w), 852 (w), 795 (w), 734 (w), 716 (w), 653 (w), 605 (w), 575 (w), 505 (w), 488 (w), 439 (vw) cm⁻¹. – MS (EI, 70 eV) *m/z* (%) = 252 (61) [M]⁺, 237 (31) [M– O]⁺, 147 (24) [M–C₈H₉]⁺, 131 (36) [M–C₈H₉O]⁺, 130 (100) [M–C₈H₁₀O]⁺, 105 (80) [C₈H₉]⁺, 104 (78) [C₈H₈]⁺. – HRMS (EI, C₁₈H₂₀O) calc. 252.1509; found 252.1510.

As no x-ray suitable crystals could be obtained, determination of the absolute configuration was done by NOESY for *(S$_P$,S)* 1-(4-[2.2]paracyclophanyl)ethanol and is in accordance with Dorizon *et al.*[172]

The *(S_P,R)* title product **141** was obtained as a white solid in 20% yield (192 mg, 731 µmol).

R_f = 0.27 (*n*-hexane/ethyl acetate; 4:1). – ¹H NMR (500 MHz, DMSO-d_6) δ = 6.54 – 6.46 (m, 3H), 6.44 – 6.38 (m, 2H), 6.34 (dd, J = 7.7, 1.6 Hz, 1H), 6.26 (s, 1H), 4.64 (p, J = 6.2 Hz, 1H, CHOH), 4.60 (d, J = 6.2 Hz, 1H, OH), 3.59 (ddd, J = 12.8, 9.5, 3.0 Hz, 1H), 3.10 – 2.88 (m, 6H), 2.79 (ddd, J = 13.2, 10.4, 6.1 Hz, 1H), 1.42 (d, J = 6.2 Hz, 3H, CH₃). – ¹³C NMR (126 MHz, DMSO-d_6) δ = 143.21 (C$_{quat}$), 139.23 (C$_{quat}$), 139.01 (C$_{quat}$), 138.96 (C$_{quat}$), 138.55 (C$_{quat}$), 134.87 (+, C$_{Ar}$H), 133.05 (+, C$_{Ar}$H), 132.84 (+, C$_{Ar}$H), 131.93 (+, C$_{Ar}$H), 131.52 (+, C$_{Ar}$H), 129.79 (+, C$_{Ar}$H), 129.69 (+, C$_{Ar}$H), 65.81 (+, CH), 34.70 (–, CH₂), 34.59 (–, CH₂), 34.06 (–, CH₂), 33.00 (–, CH₂), 21.90 (+, CH₃). – IR (ATR) ṽ = 3277 (vw), 2922 (vw), 1594 (vw), 1449 (vw), 1370 (vw), 1267 (vw), 1147 (vw), 1078 (vw), 1058 (vw), 931 (vw), 886 (vw), 850 (vw), 792 (vw), 755 (vw), 715 (vw), 652 (vw), 611 (vw), 527 (vw), 505 (vw), 462 (vw), 399 (vw) cm⁻¹. – MS (EI, 70 eV) *m/z* (%) = 252 (58) [M]⁺, 237 (31) [M–O]⁺, 147 (24) [M–C₈H₉]⁺, 131 (26) [M–C₈H₉O]⁺, 130 (100) [M–C₈H₁₀O]⁺, 105 (86) [C₈H₉]⁺, 104 (85) [C₈H₈]⁺. – HRMS (EI, C₁₈H₂₀O) calc. 252.1509; found 252.1508. The analytical data is consistent with literature.[172]

(S$_P$,S)-1-(4-[2.2]Paracyclophanyl)ethyl methyl ether (145)

In an argon purged Schlenk flask, (S$_P$,S)-1-(4-[2.2]paracyclophanyl)ethanol (100 mg, 0.40 mmol, 1.00 equiv.) was dissolved in 3 mL of dry THF and was cooled to 0 °C. To this solution, NaH (10.0 mg, 600 µmol, 1.50 equiv.) was added as a 60% suspension (w/w) in mineral oil, and the reaction mixture was stirred for 30 min at 0 °C. Methyl iodide (40.0 µL, 90.0 mg, 600 µmol, 1.50 equiv.) was then added and the mixture was stirred for another 2 h at room temperature. The reaction was quenched with sat. NH$_4$Cl solution and extracted with ethyl acetate (3 × 10 mL). The combined organic layers were washed with brine and dried over Na$_2$SO$_4$. The concentrated crude was purified by column chromatography (silica, dichloromethane/methanol; 100:1). The title compound was obtained as a white solid in 18% yield (20.2 mg, 75.9 µmol).

R$_f$ = 0.31 (dichloromethane/methanol; 100:1). − ^{1}H NMR (500 MHz, CDCl$_3$) δ = 6.61 – 6.56 (m, 2H), 6.56 – 6.49 (m, 2H), 6.44 (dd, J = 7.8, 1.8 Hz, 1H), 6.41 (dd, J = 7.7, 1.9 Hz, 1H), 6.37 (d, J = 7.6 Hz, 1H), 4.34 (q, J = 6.3 Hz, 1H, CHOMe), 3.63 (s, 3H, OCH$_3$), 3.27 (ddd, J = 13.2, 10.1, 2.5 Hz, 1H), 3.17 – 2.94 (m, 6H), 2.85 (ddd, J = 13.6, 10.6, 5.9 Hz, 1H), 1.25 (d, J = 6.3 Hz, 3H, CCH$_3$). − ^{13}C NMR (126 MHz, CDCl$_3$) δ = 143.26 (C$_{quat}$), 140.22 (C$_{quat}$), 139.73 (C$_{quat}$), 139.20 (C$_{quat}$), 135.20 (+, C$_{Ar}$H), 135.04 (C$_{quat}$), 133.20 (+, C$_{Ar}$H), 132.88 (+, C$_{Ar}$H), 132.64 (+, C$_{Ar}$H), 131.65 (+, C$_{Ar}$H), 131.01 (+, 2×C, C$_{Ar}$H), 128.59 (+, CH), 57.28 (+, OCH$_3$), 35.51 (−, CH$_2$), 35.39 (−, CH$_2$), 34.84 (−, CH$_2$), 33.22 (−, CH$_2$), 22.18 (+, CH$_3$). − IR (ATR) ṽ = 2925 (m), 2850 (w), 1735 (vw), 1592 (w), 1501 (w), 1449 (w), 1413 (w), 1365 (w), 1185 (w), 1144 (w), 1116 (m), 1088 (m), 1009 (w), 959 (w), 942 (w), 924 (w), 902 (m), 871 (w), 843 (m), 795 (m), 733 (m), 716 (m), 684 (w), 651 (m), 613 (w), 584 (w), 504 (m), 387 (vw) cm^{-1}. − MS (EI, 70 eV) m/z (%) = 267 (22) [M+H]$^+$, 266 (97) [M]$^+$, 234 (12) [M–MeOH]$^+$, 161 (32) [M–C$_8$H$_9$]$^+$, 130 (84) [C$_{10}$H$_{10}$]$^+$, 129 (100) [C$_{10}$H$_9$]$^+$, 105 (39) [C$_8$H$_9$]$^+$, 104 (61) [C$_8$H$_8$]$^+$. − HRMS (EI, C$_{19}$H$_{22}$O) calc. 266.1665; found 266.1667.

(S$_P$,R)-1-(4-[2.2]Paracyclophanyl)ethyl methyl ether (146)

In an argon purged Schlenk flask, *(S$_P$,R)*-1-(4-[2.2]paracyclophanyl)ethanol (50.0 mg, 200 μmol, 1.00 equiv.) was dissolved in 3 mL of dry THF and was cooled to 0 °C. To this solution, NaH (5.0 mg, 300 μmol, 1.50 equiv.) was added as a 60 % suspension (w/w) in mineral oil, and the reaction mixture was stirred for 30 min at 0 °C. Methyl iodide (20.0 μL, 45.0 mg, 300 μmol, 1.50 equiv.) was then added and the mixture was again stirred for 2 h at room temperature. The reaction was quenched with sat. NH$_4$Cl solution and extracted with ethyl acetate (3 × 10 mL). The combined organic layers were washed with brine and dried over Na$_2$SO$_4$. The concentrated crude was purified by column chromatography (silica, dichloromethane/methanol; 100:1). The title compound was obtained as a white solid in 92% yield (51.0 mg, 191 μmol).

R_f = 0.23 (dichloromethane/methanol; 100:1). – ^{1}H NMR (500 MHz, CDCl$_3$) δ = 6.58 – 6.53 (m, 2H), 6.49 (d, J = 7.7 Hz, 1H), 6.47 – 6.43 (m, 2H), 6.40 (s, 1H), 6.37 – 6.32 (m, 1H), 4.46 (q, J = 6.5 Hz, 1H, CHOMe), 3.49 (ddd, J = 13.1, 10.1, 2.6 Hz, 1H), 3.14 (s, 3H, OCH$_3$), 3.12 – 2.85 (m, 7H), 1.59 (d, J = 6.5 Hz, 3H, CCH$_3$). – ^{13}C NMR (126 MHz, CDCl$_3$) δ = 139.90 (C$_{quat}$), 139.72 (C$_{quat}$), 139.60 (C$_{quat}$), 139.56 (C$_{quat}$), 139.10 (C$_{quat}$), 135.20 (+, C$_{Ar}$H), 133.33 (+, C$_{Ar}$H), 133.06 (+, C$_{Ar}$H), 132.61 (+, C$_{Ar}$H), 132.10 (+, C$_{Ar}$H), 130.42 (+, C$_{Ar}$H), 130.16 (+, C$_{Ar}$H), 75.19 (+, CH), 55.13 (+, OCH$_3$), 35.48 (–, CH$_2$), 35.35 (–, CH$_2$), 35.10 (–, CH$_2$), 33.47 (+, CH$_3$), 18.05 (–, CH$_2$). – IR (ATR) $\tilde{v}$ = 2976 (w), 2922 (m), 2850 (m), 2810 (w), 1593 (w), 1499 (w), 1436 (w), 1412 (w), 1372 (w), 1343 (w), 1325 (w), 1228 (w), 1203 (w), 1145 (w), 1123 (w), 1084 (s), 1066 (m), 1015 (w), 939 (w), 925 (w), 905 (m), 863 (w), 828 (m), 793 (w), 751 (m), 716 (m), 682 (w), 650 (m), 607 (m) cm^{-1}. – MS (EI, 70 eV) *m/z* (%) = 266 (100) [M]$^+$, 234 (13) [M–MeOH]$^+$, 161 (30) [M–C$_8$H$_9$]$^+$, 130 (79) [C$_{10}$H$_{10}$]$^+$, 129 (98) [C$_{10}$H$_9$]$^+$, 105 (36) [C$_8$H$_9$]$^+$, 104 (50) [C$_8$H$_8$]$^+$. – HRMS (EI, C$_{19}$H$_{22}$O) calc. 266.1665; found 266.1664.

(*rac*)-4-Trifluoroacetyl[2.2]paracyclophane (147)

In a 500 mL round-bottomed flask trifluoracetic anhydride (13.6 mL, 20.16 g, 96.0 mmol, 2.00 equiv.) and aluminium chloride (11.2 g, 84.0 mmol, 1.75 equiv.) were suspended in 200 mL of dry dichloromethane and cooled to 0 °C. Then [2.2]paracyclophane (10.0 g, 48.0 mmol, 1.00 equiv.) was added portion-wise over 15 min. The reaction mixture was stirred for 90 min at 0 °C. Water was carefully added, the layers were separated and the organic layer was washed with 1 M HCl (200 mL), brine (200 mL) and 1 M NaOH (200 mL). The organic layer was dried over Na_2SO_4 and the solvent was removed under reduced pressure. The crude solid was purified by column chromatography (silica, cyclohexane/dichloromethane; 10:1) to yield 10.0 g of the title compound (32.9 mmol, 69%) as a white solid.

R_f = 0.39 (cyclohexane/dichloromethane; 10.1). – ^{1}H NMR (400 MHz, CDCl$_3$) δ = 7.13 (t, J = 1.8 Hz, 1H), 6.81 (dd, J = 7.8, 1.8 Hz, 1H), 6.65 (d, J = 7.8 Hz, 1H), 6.60 (dd, J = 7.9, 1.9 Hz, 1H), 6.54 (dd, J = 7.9, 1.9 Hz, 1H), 6.44 (dd, J = 7.9, 1.9 Hz, 1H), 6.39 (dd, J = 7.9, 1.9 Hz, 1H), 3.97 (ddd, J = 12.8, 7.7, 3.8 Hz, 1H), 3.30 – 3.16 (m, 4H), 3.09 – 3.00 (m, 2H), 2.99 – 2.89 (m, 1H) ppm. – ^{13}C NMR (101 MHz, CDCl$_3$) δ = 181.23 (q, J = 33.2 Hz, C$_{quat}$, CO), 145.53 (C$_{quat}$), 140.37 (C$_{quat}$), 140.03 (C$_{quat}$), 139.50 (C$_{quat}$), 138.99 (+, C$_{Ar}$H), 137.07 (+, C$_{Ar}$H), 134.77 (q, J = 3.7 Hz, +, C$_{Ar}$H), 133.14 (+, C$_{Ar}$H), 133.07 (+, C$_{Ar}$H), 132.52 (+, C$_{Ar}$H), 131.46 (+, C$_{Ar}$H), 130.17 (C$_{quat}$), 116.62 (q, J = 293.2 Hz, C$_{quat}$, CF$_3$), 36.44 (–, CH$_2$), 35.20 (–, CH$_2$), 35.17 (–, CH$_2$), 34.65 (–, CH$_2$) ppm. – ^{19}F NMR (376 MHz, CDCl$_3$) δ = –74.97 ppm. – IR (ATR) ṽ = 3043 (vw), 2925 (w), 2521 (vw), 2322 (vw), 2165 (vw), 2050 (vw), 1918 (vw), 1704 (w), 1550 (w), 1436 (w), 1288 (vw), 1200 (m), 1132 (m), 984 (w), 843 (w), 707 (w), 631 (w), 508 (w), 386 (vw), cm^{-1}. – HRMS (ESI, C$_{18}$H$_{15}$F$_3$O) calc. 205.1148 [M+H]$^+$, found 305.1130 [M+H]$^+$. The analytical data is consistent with literature.[230]

(rac)-1-(4-[2.2]Paracyclophanyl)-1-trifluoromethyl methanol (148)

In a 50 mL Schlenk flask, *(rac)*-4-trifluoroacetyl[2.2]paracyclophane (1.00 g, 3.30 mmol, 1.00 equiv.) was dissolved in 20 mL of dry tetrahydrofurane and cooled to 0 °C. Then LiAlH$_4$ (150 mg, 3.96 mmol, 1.20 equiv.) was added portion-wise. The mixture was stirred for for 1 h at this temperature, then quenched by the addition of water and acidified to pH = 6 by the addition of 1 M HCl. The phases were separated and the aqueous phase was extracted with ethyl acetate (3 × 30 mL). The combined organic layers were washed with NaHCO$_3$ solution (50 mL), dried over Na$_2$SO$_4$ and the solvents were removed under reduced pressure. The crude solid was purified by column chromatography (silica, cyclohexane/ethyl acetate; 10:1) to yield 940 mg of the title compound (3.07 mmol, 93%) as a diastereomeric mixture in the form of a white solid.

The analytical data is given for the inseparable diastereomeric mixture:

R_f = 0.26 (cyclohexane/ethyl acetate; 10:1). – ^{1}H NMR (500 MHz, CDCl$_3$) δ = 6.77 (d, J = 1.8 Hz, 1H), 6.61 – 6.45 (m, 12H), 6.39 (d, J = 8.2 Hz, 1H), 5.13 (q, J_{HF} = 6.5 Hz, 1H), 4.97 (q, J_{HF} = 7.2 Hz, 1H), 3.60 (ddd, J = 13.2, 9.7, 3.0 Hz, 1H), 3.41 (ddd, J = 14.1, 10.0, 2.1 Hz, 1H), 3.27 – 3.11 (m, 7H), 3.10 – 2.85 (m, 7H), 2.56 (bs, 1H), 2.21 (bs, 1H). – ^{13}C NMR (126 MHz, CDCl$_3$) δ = 140.48 (C$_{quat}$), 140.42 (C$_{quat}$), 139.92 (C$_{quat}$), 139.72 (C$_{quat}$), 139.66 (C$_{quat}$), 139.44 (C$_{quat}$), 139.05 (C$_{quat}$), 137.80 (C$_{quat}$), 136.61 (+, C$_{Ar}$H), 135.44 (+, C$_{Ar}$H), 134.40 (+, C$_{Ar}$H), 134.04 (+, C$_{Ar}$H), 133.61 (+, C$_{Ar}$H), 133.15 (+, C$_{Ar}$H), 132.87 (+, C$_{Ar}$H), 132.77 (+, C$_{Ar}$H), 132.70 (+, C$_{Ar}$H), 132.60 (+, C$_{Ar}$H), 132.02 (C$_{quat}$), 131.88 (C$_{quat}$), 131.47 (+, C$_{Ar}$H), 131.45 (+, C$_{Ar}$H), 131.40 (+, C$_{Ar}$H), 130.69 (+, C$_{Ar}$H), 124.99 (q, J = 282.8 Hz, C$_{quat}$, CF$_3$), 124.43 (q, J = 282.4 Hz, C$_{quat}$, CF$_3$), 72.62 (q, J = 31.4 Hz, +, CH), 69.81 (q, J = 31.4 Hz, +, CH), 35.47 (–, CH$_2$), 35.36 (–, CH$_2$), 35.33 (–, CH$_2$), 35.21 (–, 2 × CH$_2$), 34.69 (–, CH$_2$), 33.78 (–, CH$_2$), 33.46 (–, CH$_2$) ppm. – ^{19}F NMR (471 MHz, CDCl$_3$) δ = –75.00 (d, J_{FH} = 7.2 Hz), –78.36 (d, J_{FH} = 6.5 Hz) ppm. – IR (ATR) ṽ = 3573 (vw), 3428 (vw), 2926 (vw), 2854 (vw), 2704 (vw), 2636 (vw), 1902 (vw), 1595 (vw), 1499 (vw), 1411 (vw), 1262 (w), 1121 (w), 903 (vw), 800 (w), 720 (w), 639 (w), 513 (w) cm^{-1}. – HRMS (ESI, C$_{18}$H$_{17}$F$_3$O) calc. 307.1304 [M+H]$^+$, found 307.1288 [M+H]$^+$; calc. 289.1199 [M–OH]$^+$, found 289.1184 [M–OH]$^+$. The analytical data is consistent with literature.[181]

(rac)-4-[2.2]Paracyclophanecarboxamide (151)

In a 100 mL round-bottom flask, *(rac)*-4-cyano[2.2]paracyclophane (150 mg, 650 mmol, 1.00 equiv.) and K_2CO_3 (90.0 mg, 650 μmol, 1.00 equiv.) were dissolved in 30 mL of DMSO. The mixture was cooled to 0 °C and H_2O_2 (147 μL of an 30 wt% aqueous solution, 44.2 mg, 1.30 mmol, 2.00 equiv.) was carefully added. The reaction mixture was diluted with 1 M KOH solution (20 mL) and extracted with ethyl acetate (3 × 20 mL). The combined organic layers were dried over Na_2SO_4, and concentrated under reduced pressure. The crude solid was purified by short column chromatography (silica, dichloromethane/methanol; 100:1) to yield 147 mg of the title compound (584 mmol, 92%) as a white solid.

R_f = 0.19 (dichloromethane/methanol; 100:1). – ^{1}H NMR (500 MHz, DMSO-d_6) δ = 7.39 (bs, NH, 1H), 7.12 (bs, NH, 1H), 6.75 (d, J = 1.9 Hz, 1H), 6.62 – 6.56 (m, 2H), 6.55 – 6.53 (m, 2H), 6.48 (d, J = 7.7 Hz, 1H), 6.44 (d, J = 7.8 Hz, 1H), 3.74 (ddd, J = 12.5, 7.6, 4.8 Hz, 1H), 3.15 – 3.05 (m, 2H), 3.05 – 2.88 (m, 4H), 2.86 – 2.72 (m, 1H). – ^{13}C NMR (126 MHz, DMSO-d_6) δ = 170.12 (C_{quat}), 139.42 (C_{quat}), 139.28 (C_{quat}), 139.22 (C_{quat}), 139.04 (C_{quat}), 135.66 (+, $C_{Ar}H$), 135.18 (C_{quat}), 134.47 (+, $C_{Ar}H$), 132.62 (+, $C_{Ar}H$), 132.47 (+, $C_{Ar}H$), 132.44 (+, $C_{Ar}H$), 132.08 (+, $C_{Ar}H$), 131.36 (+, $C_{Ar}H$), 34.85 (–, CH_2), 34.75 (–, CH_2), 34.55 (–, CH_2), 34.50 (–, CH_2) ppm. – IR (ATR) $\tilde{v}$ = 3410 (w), 3193 (w), 3016 (vw), 2920 (w), 2849 (vw), 1681 (vw), 1628 (w), 1607 (w), 1552 (w), 1489 (vw), 1406 (w), 1376 (w), 1334 (w), 1279 (w), 1130 (w), 931 (w), 902 (w), 861 (vw), 795 (w), 761 (w), 713 (w), 615 (w), 564 (w), 515 (w), 499 (w), 455 (w), 411 (w) cm^{-1}. – MS (EI, 70 eV) *m/z* (%) = 252 (12) [M+H]$^+$, 251 (67) [M]$^+$, 147 (100) [C_9H_9NO]$^+$, 104 (46) [C_8H_8]$^+$. – HRMS (EI, $C_{17}H_{17}NO$) calc. 251.1305; found 251.1303. The analytical data is consistent with literature.[231-232]

(rac)-N-tert-Butyl-4-[2.2]paracyclophanecarboxamide (152)

 In a 25 mL round-bottomed flask, (rac)-4-cyano[2.2]paracyclophane (50 mg, 200 mmol, 1.00 equiv.) was dissolved in 5 mL of tert-butanol. Concentrated H_2SO_4 (50.0 µL, 100 mg, 1.00 mmol, 5.00 equiv.) was added and the reaction mixture was heated to reflux for 16 h. It was diluted with 1 M KOH solution (10 mL) and extracted with ethyl acetate (3 × 10 mL). The combined organic layers were washed with 1 M KOH solution (20 mL), dried over Na_2SO_4 and the solvent was removed under reduced pressure. The crude product was purified by column chromatography (silica, cyclohexane/ethyl acetate; 20:1) to yield 17.3 mg of the title compound (56.3 µmol, 28%) as a white solid.

R_f = 0.36 (cyclohexane/ethyl acetate; 20:1). – ^1H NMR (400 MHz, DMSO-d_6) δ = 7.21 (s, NH, 1H), 6.67 (d, J = 7.8 Hz, 1H), 6.62 (d, J = 1.9 Hz, 1H), 6.54 (dd, J = 7.8, 1.9 Hz, 1H), 6.53 – 6.51 (m, 2H), 6.45 (d, J = 7.7 Hz, 1H), 6.36 (d, J = 7.7 Hz, 1H), 3.56 (ddd, J = 12.6, 8.9, 3.4 Hz, 1H), 3.13 – 3.02 (m, 2H), 3.02 – 2.87 (m, 4H), 2.80 (ddd, J = 12.6, 9.8, 6.2 Hz, 1H), 1.37 (s, 9H). – ^{13}C NMR (101 MHz, DMSO-d_6) δ = 168.13 (C_{quat}, CON), 139.18 (C_{quat}), 139.15 (C_{quat}), 139.00 (C_{quat}), 138.32 (C_{quat}), 136.33 (C_{quat}), 135.39 (+, $C_{Ar}H$), 133.96 (+, $C_{Ar}H$), 132.44 (+, $C_{Ar}H$), 132.37 (+, $C_{Ar}H$), 132.31 (+, $C_{Ar}H$), 132.06 (+, $C_{Ar}H$), 131.49 (+, $C_{Ar}H$), 50.49 (C_{quat}, CMe$_3$), 34.88 (–, CH_2), 34.72 (–, CH_2), 34.52 (–, CH_2), 34.48 (–, CH_2), 28.63 (+, CH_3) ppm. – IR (ATR) ṽ = 3290 (vw), 2923 (w), 2850 (vw), 1634 (w), 1533 (w), 1449 (w), 1391 (vw), 1361 (vw), 1305 (w), 1222 (w), 906 (vw), 881 (vw), 814 (w), 717 (w), 667 (vw), 635 (w), 586 (vw), 503 (w) cm^{-1}. – MS (EI, 70 eV) m/z (%) = 308 (14) [M+H]$^+$, 307 (60) [M]$^+$, 251 (39) [$C_{17}H_{17}NO$]$^+$, 203 (9), [$C_{13}H_{17}NO$]$^+$, 147 (100) [C_9H_9NO]$^+$, 105 (11) [C_8H_9]$^+$, 104 (23) [C_8H_8]$^+$. – HRMS (EI, $C_{21}H_{25}NO$) calc. 307.1936; found 307.1936. The analytical data is consistent with literature.[233]

(rac)-N-(4-Methoxy)phenyl-4-[2.2]paracyclophanecarboxamide (154)

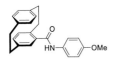

In a 50 mL round-bottomed flask, (rac)-4-carboxy[2.2]paracyclophane (250 mg, 1.00 mmol, 1.00 equiv.) was suspended in 20 mL of dry dichloromethane, cooled to 0 °C and thionyl chloride (90.0 μL, 140 mg, 1.20 mmol, 1.20 equiv.) was then added slowly. The reaction mixture was heated to reflux for 4 h until a clear solution was formed. The mixture was then concentrated in vacuo and the residue was redissolved in 10 mL of dry tetrahydrofurane at 0 °C. Dry triethylamine (210 μL, 150 mg, 1.50 mmol, 1.50 equiv.) was added and stirred for 5 min, followed by the addition of anisidine (185 mg, 1.50 mmol, 1.50 equiv.). The resulting mixture was stirred at room temperature for another 4 h and was then diluted with brine (10 mL) and extracted with ethyl acetate (3 × 10 mL). The combined organic layers were washed with brine (10 mL), dried over Na_2SO_4 and the solvents were removed under reduced pressure. The crude compound was purified by column chromatography (silica, cyclohexane/ethyl acetate; 5/1) to yield 158 mg of the title compound (442 μmol, 44%) as a white solid.

R_f = 0.41 (cyclohexane/ethyl acetate; 5:1). – ^{1}H NMR (400 MHz, DMSO-d_6) δ = 9.69 (s, NH, 1H), 7.68 (d, J = 9.0 Hz, 2H), 6.91 (d, J = 9.0 Hz, 2H), 6.79 (d, J = 1.8 Hz, 1H), 6.70 (d, J = 8.0 Hz, 1H), 6.64 (dd, J = 7.8, 1.8 Hz, 1H), 6.58 – 6.55 (m, 2H), 6.53 (d, J = 7.8 Hz, 1H), 6.41 (d, J = 7.8 Hz, 1H), 3.74 (s, 3H), 3.58 (ddd, J = 12.4, 9.7, 2.6 Hz, 1H), 3.19 – 2.83 (m, 7H) ppm. – 13C NMR (101 MHz, DMSO-d_6) δ = 166.82 (C_{quat}), 155.22 (C_{quat}), 139.50 (C_{quat}), 139.27 (C_{quat}), 139.11 (C_{quat}), 139.07 (C_{quat}), 135.63 (+, $C_{Ar}H$), 134.92 (C_{quat}), 134.79 (+, $C_{Ar}H$), 132.78 (C_{quat}), 132.48 (+, $C_{Ar}H$), 132.45 (+, 2 × $C_{Ar}H$), 132.19 (+, $C_{Ar}H$), 131.36 (+, $C_{Ar}H$), 121.29 (+, $C_{Ar}H$), 113.70 (+, $C_{Ar}H$), 55.20 (+, CH_3), 34.96 (–, CH_2), 34.68 (–, CH_2), 34.46 (–, 2 × CH_2) ppm. – IR (ATR) $\tilde{v}$ = 3233(vw), 2928 (w), 2852 (vw), 1645 (w), 1598 (w), 1505 (m), 1462 (w), 1439 (w), 1410 (w), 1313 (w), 1299 (w), 1225 (m), 1175 (w), 1108 (w), 1029 (w), 942 (w), 902 (w), 862 (vw), 830 (m), 812 (m), 717 (w), 640 (w), 591 (w), 553 (w), 512 (m), 450 (vw) cm^{-1}. – MS (EI, 70 eV) m/z (%) = 358.3 (27) [M+H]$^+$, 357 (100) [M]$^+$, 253 (83) [M–C_8H_8]$^+$, 131 (42) [C_9H_7O]$^+$, 104 (12) [C_8H_8]$^+$. – HRMS (EI, $C_{24}H_{23}NO_2$) calc. 357.1729; found 357.1729. – EA ($C_{24}H_{23}NO_2$) calc. C: 80.64, H: 6.49, N: 3.92; found C: 80.24, H: 6.48, N: 3.89. The analytical data is consistent with literature.[234]

5.6   Analytical Data of 4-Pyridyl[2.2]paracyclophane Chemo-sensor Project

(rac/S_P/R_P)-4-(4'-Pyridyl)[2.2]paracyclophane (157)

A sealable vial was charged with (rac)-4-bromo[2.2]paracyclophane (200 mg, 0.70 mmol, 1.00 equiv.), 4-pyridylboronic acid (170 mg, 1.40 mmol, 2.00 equiv.), K$_3$PO$_4$ (220 mg, 1.05 mmol, 1.50 equiv.) and Pd(PPh$_3$)$_4$ (50.0 mg, 0.04 mmol, 6.0 mol%). The sealed vial was evacuated and purged with argon three times. Then 3 mL of dioxane/water (2:1) were added *via* the septum and the reaction mixture was heated to 110 °C and stirred for 24 h. It was diluted with water (20 mL) and extracted with ethyl acetate (3 × 25 mL). The combined organic layers were dried over Na$_2$SO$_4$, filtered and the solvent was removed under reduced pressure. The residue war purified by column chromatography (silica, cyclohexane/ethyl acetate 2:1) to give 45.1 mg of the title compound (158 mg, 79%) as white solid.

R$_f$ = 0.43 (cyclohexane/ethyl acetate; 2:1). – M.p. 115–118 °C. – ^{1}H NMR (400 MHz, CDCl$_3$) δ = 8.70 (d, J = 4.8 Hz, 2H), 7.40 (d, J = 4.8 Hz, 2H), 6.65–6.62 (m, 2H), 6.59–6.56 (m, 3H), 6.54 (s, 2H), 3.40 (ddd, J = 12.5, 10.1, 2.9 Hz, 1H), 3.21–3.12 (m, 3H), 3.09–2.87 (m, 3H), 2.66 (ddd, J = 13.0, 10.0, 4.5 Hz, 1H) ppm. – ^{13}C NMR (101 MHz, CDCl$_3$) δ = 150.2 (+, C$_{Ar}$H), 148.7 (C$_{quat}$), 140.3 (C$_{quat}$), 139.7 (C$_{quat}$, 2C), 139.3 (C$_{quat}$), 137.4 (C$_{quat}$), 136.3 (+, C$_{Ar}$H), 133.6 (+, C$_{Ar}$H), 133.4 (+, C$_{Ar}$H), 132.9 (+, C$_{Ar}$H), 132.2 (+, C$_{Ar}$H), 132.1 (+, C$_{Ar}$H), 129.9 (+, C$_{Ar}$H), 124.6 (+, C$_{Ar}$H), 35.6 (–, CH$_2$), 35.4 (–, CH$_2$), 35.1 (–, CH$_2$), 34.1 (–, CH$_2$) ppm. – IR (ATR) $\tilde{v}$ = 2925 (w), 2850 (w), 1895 (w), 1594 (w), 1497 (w), 1474 (vw), 1400 (w), 1211 (vw), 1093 (w), 1066 (vw), 991 (w), 939 (vw), 905 (w), 850 (w), 825 (w), 716 (w), 667 (vw), 646 (w), 622 (w), 594 (w), 563 (w), 515 (w), 483 (w), 423 (vw) cm^{-1}. – MS (EI, 70 eV) m/z (%) = 286 (25) [M+H]$^+$, 285 (100) [M]$^+$, 181 (62) [M–C$_8$H$_7$]+, 180 (100) [M–C$_8$H$_8$]$^+$, 105 (51) [C$_8$H$_9$]$^+$, 104 (61) [C$_8$H$_8$]$^+$. – HRMS (EI, C$_{21}$H$_{19}$N) calc. 285.1512, found 285.1511.

(S$_P$/R$_P$)-*N*-Methyl-4-pyridylium[2.2]paracyclophane iodide (155)

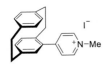

In a round-bottom flaks, *(S$_P$/R$_P$)*-4-(4'-pyridyl)[2.2]paracyclophane (50.0 mg, 180 μmol, 1.00 equiv.) was dissolved in 1.5 mL DCM. Methyl iodide (1.00 mL, 2.28 g, 16.1 mmol, 894 equiv.) was added at room temperature resulting in a yellow solution. The crude mixture was allowed to react for 3 days, while two portions of MeI (1 mL) were additionally added after each day. The reaction mixture was extracted three times with water. After the removal of water by lyophilization, 42.1 mg of the title compound (98.5 μmol, 55%) was obtained as a yellow powder.

^{1}H NMR (500 MHz, D$_2$O) δ = 8.76 (d, J = 6.4, 2H), 8.14 (d, J = 6.5, 2H), 6.86 (d, J = 11.5, 3H), 6.80 – 6.68 (m, 3H), 6.53 (d, J = 7.9, 1H), 4.38 (s, 3H), 3.45 – 3.35 (m, 1H), 3.25 – 3.06 (m, 5H), 3.03 – 2.94 (m, 1H), 2.62 – 2.52 (m, 1H) ppm. – ^{13}C NMR (126 MHz, D$_2$O) δ = 144.6 (+, CArH), 141.6 (C$_{quat}$), 140.6 (C$_{quat}$), 140.0 (C$_{quat}$), 138.8 (C$_{quat}$), 137.0 (+, CArH), 135.9 (+, CArH), 135.8 (C$_{quat}$), 133.7 (+, CArH), 133.2 (+, CArH), 132.4 (+, CArH), 131.8 (+, CArH), 129.5 (+, CArH), 127.7 (+, CArH), 47.2 (+, CH$_3$), 34.6 (–, CH$_2$), 34.3 (–, CH$_2$), 34.2 (–, CH$_2$), 33.4 (–, CH$_2$) ppm. – IR (ATR) $\bar{v}$ = 3386 (vw), 3032 (vw), 2920 (vw), 2850 (vw), 2573, (vw), 2322 (vw), 2168 (vw), 2020 (vw), 1918 (vw), 1639 (vw), 1516 (vw), 1454 (vw), 1332 (vw), 1198 (vw), 1096 (vw), 949 (vw), 843 (w), 720 (vw), 642 (vw), 488 (vw) cm^{-1}. – HRMS (ESI, C$_{22}$H$_{22}$N$^+$) calc. 300.1747, found 300.1737.

4,16-Di-(4'-pyridyl)[2.2]paracyclophane (158)

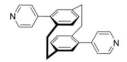

A sealable vial was charged with 4,16-dibromo[2.2]paracyclophane (260 mg, 0.70 mmol, 1.00 equiv.), 4-pyridylboronic acid (340 mg, 2.80 mmol, 4.00 equiv.), K_3PO_4 (450 mg, 2.10 mmol, 3.00 equiv.) and Pd(PPh$_3$)$_4$ (50.0 mg, 0.04 mmol, 6 mol%). The sealed vial was evacuated and purged with argon three times. Then 3 mL of dioxane/water (2:1) were added *via* the septum and the reaction mixture was heated to 110 °C and stirred for 24 h. It was diluted with water (20 mL) and extracted with ethyl acetate (3 × 25 mL). The combined organic layers were dried over Na$_2$SO$_4$, filtered and the solvent was removed under reduced pressure. The residue war purified by column chromatography (silica, ethyl acetate) to give 75.0 mg of the title compound (207 µmol, 30%) as white solid.

R_f = 0.21 (ethyl acetate). – M.p. 291–294 °C. – ^{1}H NMR (400 MHz, CDCl$_3$) δ = 8.73 (d, J = 4.4 Hz, 4H), 7.44 (d, J = 4.4 Hz, 4H), 6.70–6.67 (m, 4H), 6.59 (dd, J = 7.8, 1.9 Hz, 2H), 3.43 (ddd, J = 13.8, 10.1, 4.1 Hz, 2H), 3.08 (ddd, J = 13.8, 10.2, 4.7 Hz, 2H), 2.91 (ddd, J = 13.70, 10.2, 4.1 Hz, 2H), 2.79 (ddd, J = 13.7, 10.1, 4.7 Hz, 2H) ppm. – ^{13}C NMR (101 MHz, CDCl$_3$) δ = 150.3 (+, C$_{Ar}$H), 148.5 (C$_{quat}$), 140.3 (C$_{quat}$), 139.7 (C$_{quat}$), 137.3 (C$_{quat}$), 135.2 (+, C$_{Ar}$H), 132.4 (+, C$_{Ar}$H), 130.5 (+, C$_{Ar}$H), 124.6 (+, C$_{Ar}$H), 34.8 (–, CH$_2$), 33.7 (–, CH$_2$) ppm. – IR (ATR) $\tilde{v}$ = 2930 (w), 1731 (vw), 1589 (w), 1538 (w), 1474 (w), 1416 (w), 1400 (w), 1321 (w), 1212 (w), 1104 (w), 1065 (w), 989 (w), 913 (w), 873 (w), 821 (m), 750 (w), 733 (w), 667 (w), 654 (w), 620 (m), 571 (m), 482 (m), 423 (w) cm^{-1}. – MS (EI, 70 eV) m/z (%) = 363 (30) [M+H]$^+$, 362 (100) [M]$^+$, 181 (81) [C$_{13}$H$_{11}$N]$^+$, 180 (81) [C$_{13}$H$_{10}$N]$^+$, 104 (26) [C$_8$H$_8$]$^+$. – HRMS (EI, C$_{26}$H$_{22}$N$_2$) calc. 362.1778, found 362.1778.

(rac)-4,16-bromo-(4'-pyridyl)[2.2]paracyclophane (159)

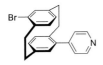

A sealable vial was charged with 4,16-dibromo[2.2]paracyclophane (260 mg, 0.70 mmol, 1.00 equiv.), 4-pyridylboronic acid (340 mg, 2.80 mmol, 4.00 equiv.), K$_3$PO$_4$ (450 mg, 2.10 mmol, 3.00 equiv.) and Pd(PPh$_3$)$_4$ (50.0 mg, 0.04 mmol, 6 mol%). The sealed vial was evacuated and purged with argon three times. Then 3 mL of dioxane/water (2:1) were added *via* the septum and the reaction mixture was heated to 110 °C and stirred for 24 h. It was diluted with water (20 mL) and extracted with ethyl acetate (3 × 25 mL). The combined organic layers were dried over Na$_2$SO$_4$, filtered and the solvent was removed under reduced pressure. The residue war purified by column chromatography (silica, ethyl acetate) to give 71.0 mg of the title compound (195 μmol, 28%) as white solid.

R$_f$ = 0.48 (ethyl acetate). – M.p. 208–211 °C. – ^{1}H NMR (400 MHz, CDCl$_3$) δ = 8.70 (d, J = 4.4 Hz, 2H), 7.38 (d, J = 4.5 Hz, 2H), 7.23 (dd, J = 7.8, 2.0 Hz, 1H), 6.64–6.59 (m, 3H), 6.53 (d, J = 7.9 Hz, 1H), 6.49 (dd, J = 7.8, 1.7 Hz, 1H), 3.56 (ddd, J = 13.0, 10.3, 2.3 Hz, 1H), 3.39–3.24 (m, 2H), 3.10–2.99 (m, 2H), 2.92 (ddd, J = 13.3, 10.6, 5.6 Hz, 1H), 2.84–2.64 (m, 2H) ppm. – ^{13}C NMR (101 MHz, CDCl$_3$) δ = 150.2 (+, C$_{Ar}$H), 148.6 (C$_{quat}$), 141.7 (C$_{quat}$), 140.1 (C$_{quat}$), 139.8 (C$_{quat}$), 139.1 (C$_{quat}$), 137.6 (+, C$_{Ar}$H), 136.9 (C$_{quat}$), 135.5 (+, C$_{Ar}$H), 134.0 (+, C$_{Ar}$H), 132.2 (+, C$_{Ar}$H), 129.8 (+, C$_{Ar}$H), 129.0 (+, C$_{Ar}$H), 126.8 (C$_{quat}$), 124.6 (+, C$_{Ar}$H), 35.8 (–, CH$_2$), 34.5 (–, CH$_2$), 33.5 (–, CH$_2$), 33.3 (–, CH$_2$) ppm. – IR (ATR) $\tilde{v}$ = 2920 (vw), 1592 (w), 1537 (w), 1474 (w), 1433 (w), 1400 (w), 1213 (w), 1130 (vw), 1067 (vw), 1031 (w), 991 (w), 958 (vw), 907 (w), 862 (w), 835 (w), 819 (w), 710 (w), 692 (w), 669 (w), 650 (w), 624 (w), 610 (w), 568 (w), 526 (vw), 484 (w), 470 (w), 431 (vw), 402 (vw) cm^{-1}. – MS (EI, 70 eV) m/z (%) = 365 (44) [M(^{81}Br)]$^+$, 363 (44) [M(^{79}Br)]$^+$, 181 (51) [C$_{13}$H$_{11}$N]$^+$, 180 (100) [C$_{13}$H$_{10}$N]$^+$. – HRMS (EI, C$_{21}$H$_{18}$N^{79}Br) calc. 363.0617, found 363.0617.

5.7   Crystal Structures

5.7.1   Crystallographic Data Solved by Dr. Martin Nieger

Crystal structures in this section were measured and solved by Dr. Martin Nieger at the University of Helsinki.

Table 25. Overview over the numbering and sample coding of crystals from Dr. Nieger.

Numbering in this thesis	Sample Code used by Dr. Nieger
87	SB829_hy
132a	SB1052_d
133	SB1034_hy
132k	SB1126_hy
132l	SB1177_hy
141	SB994_hy
146	SB995_hy
154	SB1033
157	SB860_hy
158	SB861_hy
159	SB862_hy

(rac)-4-Bromo-13-benzoyl[2.2]paracyclophane (87) SB829_hy

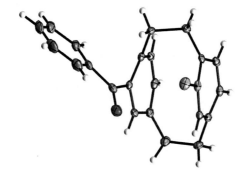

Crystal data

$C_{23}H_{19}BrO·1/6(C_3H_6O)·1/6(H_2O)$	$D_x = 1.474$ Mg m^{-3}
$M_r = 403.97$	Cu $K\alpha$ radiation, $\lambda = 1.54178$ Å
Hexagonal, $P6_3$ (no.170)	Cell parameters from 9860 reflections
$a = 20.5982$ (6) Å	$\theta = 2.5$–$72.2°$
$c = 7.4331$ (2) Å	$\mu = 3.15$ mm^{-1}
$V = 2731.23$ (18) Å^3	$T = 123$ K
$Z = 6$	Needles, colourless
$F(000) = 1242$	$0.16 \times 0.04 \times 0.02$ mm

Data collection

Bruker D8 VENTURE diffractometer with Photon100 detector	3586 independent reflections
Radiation source: INCOATEC microfocus sealed tube	3491 reflections with $I > 2\sigma(I)$
Detector resolution: 10.4167 pixels mm^{-1}	$R_{int} = 0.045$
rotation in ϕ and ω, 0.5°, shutterless scans	$\theta_{max} = 72.2°$, $\theta_{min} = 2.5°$
Absorption correction: multi-scan *SADABS* V2014/5 (Bruker AXS Inc.)	$h = -25 \rightarrow 25$
$T_{min} = 0.745$, $T_{max} = 0.929$	$k = -25 \rightarrow 25$
47487 measured reflections	$l = -9 \rightarrow 9$

Refinement

Refinement on F^2	Secondary atom site location: difference Fourier map
Least-squares matrix: fulldircet	Hydrogen site location: difference Fourier map
$R[F^2 > 2\sigma(F^2)] = 0.020$	H-atom parameters constrained
$wR(F^2) = 0.049$	$w = 1/[\sigma^2(F_o^2) + (0.0306P)^2 + 0.5064P]$ where $P = (F_o^2 + 2F_c^2)/3$
$S = 1.06$	$(\Delta/\sigma)_{max} = 0.005$
3586 reflections	$\Delta\rangle_{max} = 0.22$ e Å^{-3}
226 parameters	$\Delta\rangle_{min} = -0.21$ e Å^{-3}
1 restraint	Absolute structure: Flack x determined using 1556 quotients [(I+)-(I-)]/[(I+)+(I-)] (Parsons, Flack and Wagner, Acta Cryst. B69 (2013) 249-259). Determination of absolute structure using Bayesian statistics on Bijvoet differences (Hooft, Straver and Spek, 2008): Hooft's y = -0.026(4) (*PLATON*: Spek, 2009).
Primary atom site location: structure-invariant direct methods	Absolute structure parameter: -0.021 (7)

(rac)-4-N-Phenylamino-7-phenyl[2.2]paracyclophane (132a) SB1052_d

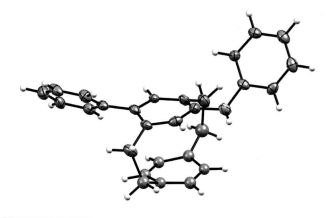

$C_{28}H_{25}N$	$F(000) = 800$
$M_r = 375.49$	$D_x = 1.231$ Mg m^{-3}
Monoclinic, $P2_1/c$ *(no.14)*	Cu $K\alpha$ radiation, $\lambda = 1.54178$ Å
$a = 12.8987$ (8) Å	Cell parameters from 6242 reflections
$b = 9.8224$ (6) Å	$\theta = 3.6–72.4°$
$c = 16.9910$ (9) Å	$\mu = 0.53$ mm^{-1}
$\beta = 109.710$ (4)°	$T = 123$ K
$V = 2026.6$ (2) Å^3	Blocks, yellow
$Z = 4$	$0.16 \times 0.12 \times 0.03$ mm

Data collection

Bruker D8 VENTURE diffractometer with Photon100 detector	3939 independent reflections
Radiation source: INCOATEC microfocus sealed tube	2892 reflections with $I > 2\sigma(I)$
Detector resolution: 10.4167 pixels mm^{-1}	$R_{int} = 0.086$
rotation in ϕ and ω, 1°, shutterless scans	$\theta_{max} = 72.5°$, $\theta_{min} = 3.6°$
Absorption correction: multi-scan *SADABS* (Sheldrick, 2014)	$h = -15 \rightarrow 15$
$T_{min} = 0.830$, $T_{max} = 0.971$	$k = -12 \rightarrow 12$
18009 measured reflections	$l = -18 \rightarrow 20$

Refinement

Refinement on F^2	Primary atom site location: dual
Least-squares matrix: full	Secondary atom site location: difference Fourier map
$R[F^2 > 2\sigma(F^2)] = 0.115$	Hydrogen site location: mixed
$wR(F^2) = 0.232$	H atoms treated by a mixture of independent and constrained refinement
$S = 1.11$	$w = 1/[\sigma^2(F_o^2) + (0.031P)^2 + 7.220P]$ where $P = (F_o^2 + 2F_c^2)/3$
3939 reflections	$(\Delta/\sigma)_{max} < 0.001$
250 parameters	$\Delta\rangle_{max} = 0.49$ e Å^{-3}
64 restraints	$\Delta\rangle_{min} = -0.46$ e Å^{-3}

(rac)-4-N,N-Diphenylamino-7-phenyl[2.2]paracyclophane (133) SB1034_hy

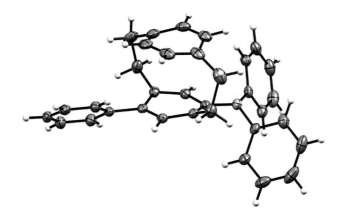

C$_{34}$H$_{29}$N	$F(000) = 960$
$M_r = 451.58$	$D_x = 1.240$ Mg m^{-3}
Monoclinic, $P2_1/c$ (no.14)	Cu $K\alpha$ radiation, $\lambda = 1.54178$ Å
$a = 8.0629$ (2) Å	Cell parameters from 7639 reflections
$b = 14.9366$ (5) Å	$\theta = 3.6–72.1°$
$c = 20.2733$ (6) Å	$\mu = 0.54$ mm^{-1}
$\beta = 97.692$ (2)°	$T = 123$ K
$V = 2419.59$ (12) Å^3	Plates, colourless
$Z = 4$	0.16 × 0.12 × 0.03 mm

Data collection

Bruker D8 VENTURE diffractometer with Photon100 detector	4719 independent reflections
Radiation source: INCOATEC microfocus sealed tube	3706 reflections with $I > 2\sigma(I)$
Detector resolution: 10.4167 pixels mm^{-1}	$R_{int} = 0.048$
rotation in ϕ and ω, 1°, shutterless scans	$\theta_{max} = 72.3°$, $\theta_{min} = 3.7°$
Absorption correction: multi-scan SADABS (Sheldrick, 2014)	$h = -9 \rightarrow 9$
$T_{min} = 0.857$, $T_{max} = 0.971$	$k = -18 \rightarrow 16$
16489 measured reflections	$l = -23 \rightarrow 25$

Refinement

Refinement on F^2	Primary atom site location: dual
Least-squares matrix: full	Secondary atom site location: difference Fourier map
$R[F^2 > 2\sigma(F^2)] = 0.057$	Hydrogen site location: difference Fourier map
$wR(F^2) = 0.126$	H-atom parameters constrained
$S = 1.10$	$w = 1/[\sigma^2(F_o^2) + (0.0397P)^2 + 1.4223P]$ where $P = (F_o^2 + 2F_c^2)/3$
4719 reflections	$(\Delta/\sigma)_{max} < 0.001$
316 parameters	$\Delta\rangle_{max} = 0.23$ e Å^{-3}
0 restraints	$\Delta\rangle_{min} = -0.25$ e Å^{-3}

(rac)-4-N-Phenylamino-7-(4-methoxy)phenyl[2.2]paracyclophane (132k) SB1126_hy

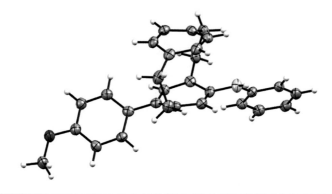

$C_{29}H_{27}NO$	$F(000) = 1728$
$M_r = 405.51$	$D_x = 1.267$ Mg m^{-3}
Monoclinic, $P2_1/c$ *(no.14)*	Cu $K\alpha$ radiation, $\lambda = 1.54178$ Å
$a = 25.015$ (1) Å	Cell parameters from 9891 reflections
$b = 10.2604$ (4) Å	$\theta = 4.6–71.9°$
$c = 17.4735$ (7) Å	$\mu = 0.58$ mm^{-1}
$\beta = 108.614$ (2)°	$T = 123$ K
$V = 4250.2$ (3) Å^3	Blocks, colourless
$Z = 8$	$0.12 \times 0.08 \times 0.04$ mm

Data collection

Bruker D8 VENTURE diffractometer with PhotonII CPAD detector	6623 reflections with $I > 2\sigma(I)$
Radiation source: INCOATEC microfocus sealed tube	$R_{int} = 0.074$
rotation in ϕ and ω, 1°, shutterless scans	$\theta_{max} = 72.9°$, $\theta_{min} = 3.7°$
Absorption correction: multi-scan *SADABS* (Sheldrick, 2014)	$h = -29 \rightarrow 30$
$T_{min} = 0.854$, $T_{max} = 0.971$	$k = -12 \rightarrow 11$
43462 measured reflections	$l = -19 \rightarrow 21$
8367 independent reflections	

Refinement

Refinement on F^2	Primary atom site location: dual
Least-squares matrix: full	Secondary atom site location: difference Fourier map
$R[F^2 > 2\sigma(F^2)] = 0.069$	Hydrogen site location: mixed
$wR(F^2) = 0.188$	H atoms treated by a mixture of independent and constrained refinement
$S = 1.03$	$w = 1/[\sigma^2(F_o^2) + (0.0913P)^2 + 3.3808P]$ where $P = (F_o^2 + 2F_c^2)/3$
8367 reflections	$(\Delta/\sigma)_{max} < 0.001$
567 parameters	$\Delta\rangle_{max} = 0.69$ e Å^{-3}
2 restraints	$\Delta\rangle_{min} = -0.28$ e Å^{-3}

(rac)-4-N-Phenylamino-7-(2-methoxy)phenyl[2.2]paracyclophane (132l) SB1177_hy

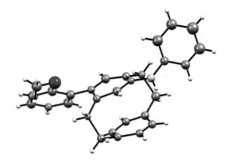

Crystal data

C$_{29}$H$_{27}$NO	Z = 8
Mr = 405.51	F(000) = 1728
Orthorhombic, Pna21 (no.33)	Dx = 1.251 Mg m-3
a = 18.3783 (7) Å	Cu Ka radiation, l = 1.54178 Å
b = 7.5162 (3) Å	m = 0.58 mm-1
c = 31.1693 (11) Å	T = 123 K
V = 4305.6 (3) Å3	0.16 × 0.08 × 0.04 mm

Due to bad data, only the constitution and conformation could be determined.

(S_P,R) 1-(4-[2.2]Paracyclophanyl)ethanol (141) SB994_hy

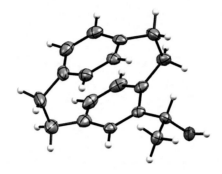

$C_{18}H_{20}O$	$D_x = 1.212$ Mg m^{-3}
$M_r = 252.34$	Cu $K\alpha$ radiation, $\lambda = 1.54178$ Å
Orthorhombic, $P2_12_12_1$ (no.19)	Cell parameters from 6417 reflections
$a = 8.0109$ (2) Å	$\theta = 3.2–71.9°$
$b = 12.8199$ (4) Å	$\mu = 0.56$ mm^{-1}
$c = 26.9361$ (9) Å	$T = 123$ K
$V = 2766.31$ (14) Å^3	Needles, colourless
$Z = 8$	$0.20 \times 0.04 \times 0.02$ mm
$F(000) = 1088$	

Data collection

Bruker D8 VENTURE diffractometer with Photon100 detector	5380 independent reflections
Radiation source: INCOATEC microfocus sealed tube	4429 reflections with $I > 2\sigma(I)$
Detector resolution: 10.4167 pixels mm^{-1}	$R_{int} = 0.067$
rotation in ϕ and ω, 1°, shutterless scans	$\theta_{max} = 72.2°$, $\theta_{min} = 3.3°$
Absorption correction: multi-scan $SADABS$ (Sheldrick, 2014)	$h = -7{\to}9$
$T_{min} = 0.874$, $T_{max} = 0.971$	$k = -15{\to}15$
18082 measured reflections	$l = -33{\to}32$

Refinement

Refinement on F^2	Secondary atom site location: difference Fourier map
Least-squares matrix: full	Hydrogen site location: difference Fourier map
$R[F^2 > 2\sigma(F^2)] = 0.087$	H atoms treated by a mixture of independent and constrained refinement
$wR(F^2) = 0.229$	$w = 1/[\sigma^2(F_o^2) + (0.1489P)^2 + 0.3337P$, where $P = (F_o^2 + 2F_c^2)/3$
$S = 1.09$	$(\Delta/\sigma)_{max} < 0.001$
5380 reflections	$\Delta\rangle_{max} = 0.31$ e Å^{-3}
349 parameters	$\Delta\rangle_{min} = -0.44$ e Å^{-3}
2 restraints	Absolute structure: Flack x determined using 1521 quotients [(I+)-(I-)]/[(I+)+(I-)] (Parsons, Flack and Wagner, Acta Cryst. B69 (2013) 249-259). The absolute configuration has not been established by anomalous dispersion effects in diffraction measurement on the crystal. The enantiomer has been assigned by reference to an unchanging stereogenic centre in the synthetic procedure (Sp).
Primary atom site location: structure-invariant direct methods	Absolute structure parameter: -0.2 (3)

(S$_P$,R) 1-(4-[2.2]Paracyclophanyl)ethyl methyl ether (146) SB995_hy

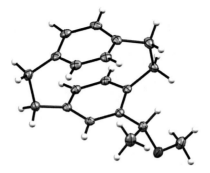

C$_{19}$H$_{22}$O	D_x = 1.247 Mg m^{-3}
M_r = 266.36	Cu $K\alpha$ radiation, λ = 1.54178 Å
Orthorhombic, $P2_12_12_1$ (no.19)	Cell parameters from 7908 reflections
a = 7.5361 (3) Å	θ = 4.7–71.8°
b = 11.1070 (4) Å	μ = 0.57 mm^{-1}
c = 16.9547 (6) Å	T = 123 K
V = 1419.17 (9) Å^3	Plates, colourless
Z = 4	0.20 × 0.10 × 0.02 mm
$F(000)$ = 576	

Data collection

Bruker D8 VENTURE diffractometer with Photon100 detector	2759 independent reflections
Radiation source: INCOATEC microfocus sealed tube	2642 reflections with $I > 2\sigma(I)$
Detector resolution: 10.4167 pixels mm^{-1}	R_{int} = 0.037
rotation in ϕ and ω, 1°, shutterless scans	θ_{max} = 72.0°, θ_{min} = 4.8°
Absorption correction: multi-scan *SADABS* (Sheldrick, 2014)	h = -9→8
T_{min} = 0.806, T_{max} = 0.971	k = -13→13
9466 measured reflections	l = -18→20

Refinement

Refinement on F^2	Secondary atom site location: difference Fourier map
Least-squares matrix: full	Hydrogen site location: difference Fourier map
$R[F^2 > 2\sigma(F^2)]$ = 0.040	H-atom parameters constrained
$wR(F^2)$ = 0.103	$w = 1/[\sigma^2(F_o^2) + (0.0517P)^2 + 0.4666P]$, where $P = (F_o^2 + 2F_c^2)/3$
S = 1.08	$(\Delta/\sigma)_{max} < 0.001$
2759 reflections	$\Delta\rangle_{max}$ = 0.23 e Å^{-3}
182 parameters	$\Delta\rangle_{min}$ = -0.17 e Å^{-3}
0 restraints	Absolute structure: Flack x determined using 1061 quotients [(I+)-(I-)]/[(I+)+(I-)] (Parsons, Flack and Wagner, Acta Cryst. B69 (2013) 249-259). The absolute configuration has not been established by anomalous dispersion effects in diffraction measurement on the crystal. The enantiomer has been assigned by reference to an unchanging stereogenic centre in the synthetic procedure (Sp).
Primary atom site location: structure-invariant direct methods	Absolute structure parameter: -0.02 (19)

(rac)-N-(4-Methoxy)phenyl-4-[2.2]paracyclophanecarboxamide (154) SB1033

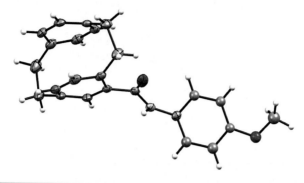

$C_{24}H_{23}NO_2$	$F(000) = 760$
$M_r = 357.43$	$D_x = 1.266$ Mg m^{-3}
Monoclinic, $P2_1/c$ *(no.14)*	Cu $K\alpha$ radiation, $\lambda = 1.54178$ Å
$a = 7.4340$ (3) Å	Cell parameters from 7815 reflections
$b = 26.5500$ (11) Å	$\theta = 3.3–72.3°$
$c = 9.6556$ (4) Å	$\mu = 0.63$ mm^{-1}
$\beta = 100.292$ (2)°	$T = 123$ K
$V = 1875.09$ (13) Å^3	Plates, colourless
$Z = 4$	$0.36 \times 0.09 \times 0.03$ mm

Data collection

Bruker D8 VENTURE diffractometer with Photon100 detector	3682 independent reflections
Radiation source: INCOATEC microfocus sealed tube	2934 reflections with $I > 2\sigma(I)$
Detector resolution: 10.4167 pixels mm^{-1}	$R_{int} = 0.042$
rotation in ϕ and ω, n°, shutterless scans	$\theta_{max} = 72.3°$, $\theta_{min} = 3.3°$
Absorption correction: multi-scan *SADABS* (Sheldrick, 2014)	$h = -9\rightarrow9$
$T_{min} = 0.845$, $T_{max} = 0.971$	$k = -32\rightarrow32$
17275 measured reflections	$l = -11\rightarrow11$

Refinement

Refinement on F^2	Primary atom site location: dual
Least-squares matrix: full	Secondary atom site location: difference Fourier map
$R[F^2 > 2\sigma(F^2)] = 0.057$	Hydrogen site location: mixed
$wR(F^2) = 0.140$	H-atom parameters constrained
$S = 1.03$	$w = 1/[\sigma^2(F_o^2) + (0.0499P)^2 + 1.7578P]$ where $P = (F_o^2 + 2F_c^2)/3$
3682 reflections	$(\Delta/\sigma)_{max} < 0.001$
237 parameters	$\Delta\rangle_{max} = 0.43$ e Å^{-3}
138 restraints	$\Delta\rangle_{min} = -0.32$ e Å^{-3}

(rac)-4-(4'-Pyridyl)[2.2]paracyclophane (157) SB860_hy

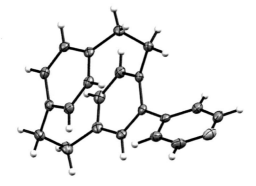

C$_{21}$H$_{19}$N	D_x = 1.281 Mg m^{-3}
M_r = 285.37	Cu $K\alpha$ radiation, λ = 1.54178 Å
Orthorhombic, *Pbca (no.61)*	Cell parameters from 9042 reflections
a = 12.7181 (7) Å	θ = 4.5–72.1°
b = 7.5633 (4) Å	μ = 0.56 mm^{-1}
c = 30.7704 (18) Å	T = 123 K
V = 2959.8 (3) Å^3	Plates, colourless
Z = 8	0.21 × 0.12 × 0.03 mm
$F(000)$ = 1216	
CCDC	1452894

Data collection

Bruker D8 VENTURE diffractometer with Photon100 detector	2927 independent reflections
Radiation source: INCOATEC microfocus sealed tube	2656 reflections with $I > 2\sigma(I)$
Detector resolution: 10.4167 pixels mm^{-1}	R_{int} = 0.036
rotation in ϕ and ω, 0.5°, shutterless scans	θ_{max} = 72.3°, θ_{min} = 2.9°
Absorption correction: multi-scan *SADABS* (Sheldrick, 2014)	h = -15→15
T_{min} = 0.895, T_{max} = 0.977	k = -9→9
30786 measured reflections	l = -37→38

Refinement

Refinement on F^2	Primary atom site location: structure-invariant direct methods
Least-squares matrix: full	Secondary atom site location: difference Fourier map
$R[F^2 > 2\sigma(F^2)]$ = 0.037	Hydrogen site location: difference Fourier map
$wR(F^2)$ = 0.104	H-atom parameters constrained
S = 1.05	$w = 1/[\sigma^2(F_o^2) + (0.0562P)^2 + 1.1212P]$ where $P = (F_o^2 + 2F_c^2)/3$
2927 reflections	$(\Delta/\sigma)_{max}$ < 0.001
199 parameters	$\Delta\rangle_{max}$ = 0.26 e Å^{-3}
0 restraints	$\Delta\rangle_{min}$ = -0.18 e Å^{-3}

(rac)-4,16-Di-(4'-pyridyl)[2.2]paracyclophane (158) SB861_hy

$C_{26}H_{22}N_2$	$F(000) = 384$
$M_r = 362.45$	$D_x = 1.336$ Mg m^{-3}
Monoclinic, $P2_1/c$ (*no.14*)	Cu $K\alpha$ radiation, $\lambda = 1.54178$ Å
$a = 10.2911$ (6) Å	Cell parameters from 9901 reflections
$b = 7.5942$ (4) Å	$\theta = 4.4–72.1°$
$c = 11.8277$ (7) Å	$\mu = 0.60$ mm^{-1}
$\beta = 102.935$ (1)°	$T = 123$ K
$V = 900.91$ (9) Å^3	Blocks, colourless
$Z = 2$	$0.30 \times 0.25 \times 0.15$ mm
CCDC	1452895

Data collection

Bruker D8 VENTURE diffractometer with Photon100 detector	1779 independent reflections
Radiation source: INCOATEC microfocus sealed tube	1742 reflections with $I > 2\sigma(I)$
Detector resolution: 10.4167 pixels mm^{-1}	$R_{int} = 0.021$
rotation in ϕ and ω, 1°, shutterless scans	$\theta_{max} = 72.2°$, $\theta_{min} = 4.4°$
Absorption correction: multi-scan *SADABS* (Sheldrick, 2014)	$h = -12{\rightarrow}12$
$T_{min} = 0.800$, $T_{max} = 0.915$	$k = -9{\rightarrow}9$
12317 measured reflections	$l = -14{\rightarrow}14$

Refinement

Refinement on F^2	Secondary atom site location: difference Fourier map
Least-squares matrix: full	Hydrogen site location: difference Fourier map
$R[F^2 > 2\sigma(F^2)] = 0.039$	H-atom parameters constrained
$wR(F^2) = 0.101$	$w = 1/[\sigma^2(F_o^2) + (0.0503P)^2 + 0.431P]$ where $P = (F_o^2 + 2F_c^2)/3$
$S = 1.04$	$(\Delta/\sigma)_{max} = 0.001$
1779 reflections	$\Delta\rangle_{max} = 0.26$ e Å^{-3}
128 parameters	$\Delta\rangle_{min} = -0.19$ e Å^{-3}
0 restraints	Extinction correction: *SHELXL2014/7* (Sheldrick 2014, Fc*=kFc[1+0.001xFc$^2\lambda^3$/sin(2θ)]$^{-1/4}$
Primary atom site location: structure-invariant direct methods	Extinction coefficient: 0.0104 (14)

(rac)-4,16-Bromo-(4'-pyridyl)[2.2]paracyclophane (159) SB862_hy

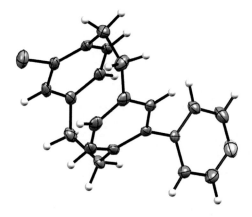

$C_{21}H_{18}BrN$	$F(000) = 744$
$M_r = 364.27$	$D_x = 1.485$ Mg m^{-3}
Monoclinic, $P2_1/n$ *(no.14)*	Cu $K\alpha$ radiation, $\lambda = 1.54178$ Å
$a = 11.2562$ (4) Å	Cell parameters from 9911 reflections
$b = 7.7962$ (3) Å	$\theta = 4.4–72.2°$
$c = 18.7312$ (7) Å	$\mu = 3.41$ mm^{-1}
$\beta = 97.494$ (1)°	$T = 123$ K
$V = 1629.73$ (10) Å^3	Plates, colourless
$Z = 4$	$0.40 \times 0.30 \times 0.02$ mm
CCDC	1452896

Data collection

Bruker D8 VENTURE diffractometer with Photon100 detector	3196 independent reflections
Radiation source: INCOATEC microfocus sealed tube	3040 reflections with $I > 2\sigma(I)$
Detector resolution: 10.4167 pixels mm^{-1}	$R_{int} = 0.041$
rotation in ϕ and ω, 1°, shutterless scans	$\theta_{max} = 72.2°$, $\theta_{min} = 4.4°$
Absorption correction: multi-scan *SADABS* (Sheldrick, 2014)	$h = -13 \rightarrow 13$
$T_{min} = 0.642$, $T_{max} = 0.929$	$k = -9 \rightarrow 9$
20127 measured reflections	$l = -20 \rightarrow 23$

Refinement

Refinement on F^2	Primary atom site location: structure-invariant direct methods
Least-squares matrix: full	Secondary atom site location: difference Fourier map
$R[F^2 > 2\sigma(F^2)] = 0.032$	Hydrogen site location: difference Fourier map
$wR(F^2) = 0.088$	H-atom parameters constrained
$S = 1.06$	$w = 1/[\sigma^2(F_o^2) + (0.0441P)^2 + 1.3326P]$ where $P = (F_o^2 + 2F_c^2)/3$
3196 reflections	$(\Delta/\sigma)_{max} = 0.001$
208 parameters	$\Delta\rangle_{max} = 0.50$ e Å^{-3}
0 restraints	$\Delta\rangle_{min} = -0.47$ e Å^{-3}

5.7.2   Crystallographic Data Solved by Dr. Olaf Fuhr

Crystal structures in this section were measured and solved by Dr. Olaf Fuhr at the Institute of Nanotechnology (INT) at the Karlsruhe Institute of Technology.

Table 26. Overview over the numbering and sample coding of crystals from Dr. Fuhr.

Numbering in this thesis	Sample Code used by Dr. Fuhr
101	ESP393
103	ESP436
113	ESP386
114	ESP465
135	ESP509_tw
76	ESP496
117	ESP-CMM18

(rac)-4-Bromo-13-carbomethoxy[2.2]paracyclophane (101) ESP393

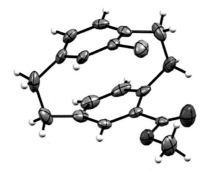

Identification code	ESP393
Empirical formula	$C_{18}H_{17}BrO_2$
Formula weight	345.22
Temperature/K	180.15
Crystal system	orthorhombic
Space group	Pna21
a/Å	17.6572(6)
b/Å	7.5893(4)
c/Å	11.1178(4)
α/°	90
β/°	90
γ/°	90
Volume/Å3	1489.85(11)
Z	4
ρcalcg/cm3	1.539
μ/mm-1	2.761
F(000)	704.0
Crystal size/mm3	0.51 × 0.28 × 0.25
Radiation	MoKα (λ = 0.71073)
2Θ range for data collection/°	4.614 to 53.52
Index ranges	-22 ≤ h ≤ 22, -9 ≤ k ≤ 9, -14 ≤ l ≤ 14
Reflections collected	10578
Independent reflections	3135 [Rint = 0.0442, Rsigma = 0.0283]
Indep. refl. with I>=2σ (I)	2819
Data/restraints/parameters	3135/1/191
Goodness-of-fit on F2	1.074
Final R indexes [I>=2σ (I)]	R1 = 0.0376, wR2 = 0.0953
Final R indexes [all data]	R1 = 0.0440, wR2 = 0.0992
Largest diff. peak/hole / e Å-3	0.43/-0.41
Flack parameter	0.004(9)

(rac)-4-Bromo-13-cyano[2.2]paracyclophane (103) ESP436

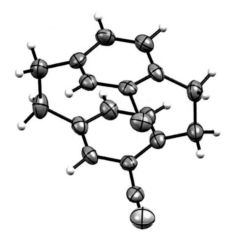

Identification code	ESB436
Empirical formula	C₁₇H₁₄BrN
Formula weight	312.20
Temperature/K	200.15
Crystal system	monoclinic
Space group	Cc
a/Å	15.0251(13)
b/Å	7.7852(7)
c/Å	11.3086(13)
α/°	90
β/°	91.030(8)
γ/°	90
Volume/Å³	1322.6(2)
Z	4
ρ_calcg/cm³	1.568
μ/mm⁻¹	2.810
F(000)	632.0
Crystal size/mm³	0.25 × 0.03 × 0.02
Radiation	GaKα (λ = 1.34143)
2Θ range for data collection/°	10.246 to 109.92
Index ranges	-14 ≤ h ≤ 18, -9 ≤ k ≤ 8, -13 ≤ l ≤ 11
Reflections collected	5831
Independent reflections	1938 [R_int = 0.0379, R_sigma = 0.0330]
Indep. refl. with I>=2σ (I)	1713
Data/restraints/parameters	1938/2/176
Goodness-of-fit on F²	1.091
Final R indexes [I>=2σ (I)]	R₁ = 0.0415, wR₂ = 0.1092
Final R indexes [all data]	R₁ = 0.0484, wR₂ = 0.1139
Largest diff. peak/hole / e Å⁻³	0.39/-0.32
Flack parameter	0.02(5)

(rac)-4-Bromo-16-cyano[2.2]paracyclophane (113) ESP386

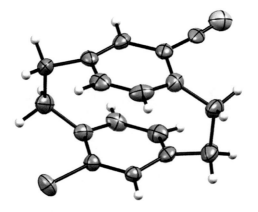

Identification code	ESP386
Empirical formula	$C_{17}H_{14}BrN$
Formula weight	312.20
Temperature/K	200.15
Crystal system	monoclinic
Space group	$P2_1$
a/Å	8.1932(6)
b/Å	7.7192(3)
c/Å	10.7451(7)
α/°	90
β/°	104.005(5)
γ/°	90
Volume/Å^3	659.37(7)
Z	2
ρ_{calc}g/cm^3	1.572
μ/mm^{-1}	3.101
F(000)	316.0
Crystal size/mm^3	0.25 × 0.21 × 0.04
Radiation	MoKα (λ = 0.71073)
2Θ range for data collection/°	3.906 to 66.134
Index ranges	$-12 \leq h \leq 12, -11 \leq k \leq 9, -11 \leq l \leq 16$
Reflections collected	9036
Independent reflections	4301 [R_{int} = 0.0243, R_{sigma} = 0.0387]
Indep. refl. with I>=2σ (I)	3376
Data/restraints/parameters	4301/1/177
Goodness-of-fit on F^2	1.111
Final R indexes [I>=2σ (I)]	R_1 = 0.0808, wR_2 = 0.2060
Final R indexes [all data]	R_1 = 0.0993, wR_2 = 0.2150
Largest diff. peak/hole / e Å^{-3}	0.72/-1.34
Flack parameter	0.27(4)

(rac)-4-Cyano-16-(4'-N,N-diphenylamino)phenyl[2.2]paracyclophane (114) ESP465

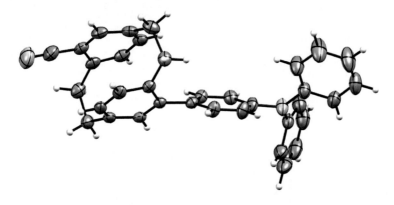

Identification code	ESP465
Empirical formula	$C_{35}H_{28}N_2$
Formula weight	476.59
Temperature/K	150.15
Crystal system	monoclinic
Space group	$P2_1$
a/Å	10.9652(4)
b/Å	7.7536(2)
c/Å	15.6511(6)
α/°	90
β/°	98.902(3)
γ/°	90
Volume/Å³	1314.62(8)
Z	2
ρ_{calc}g/cm³	1.204
μ/mm⁻¹	0.344
F(000)	504.0
Crystal size/mm³	0.19 × 0.04 × 0.03
Radiation	GaKα (λ = 1.34143)
2Θ range for data collection/°	19.984 to 104.188
Index ranges	-12 ≤ h ≤ 11, -3 ≤ k ≤ 9, -18 ≤ l ≤ 18
Reflections collected	11234
Independent reflections	3234 [R_{int} = 0.0178, R_{sigma} = 0.0150]
Indep. refl. with I>=2σ (I)	2994
Data/restraints/parameters	3234/1/334
Goodness-of-fit on F^2	1.042
Final R indexes [I>=2σ (I)]	R_1 = 0.0304, wR_2 = 0.0797
Final R indexes [all data]	R_1 = 0.0335, wR_2 = 0.0814
Largest diff. peak/hole / e Å⁻³	0.11/-0.15
Flack parameter	-1.2(8)

(rac)-[2]Paracyclo[2]6-(tert-butyl)(1,7)carbazolophane (135) ESP509_tw

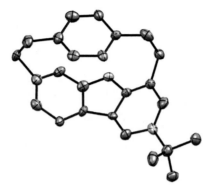

Identification code	ESP509_tw
Empirical formula	$C_{26}H_{27}N$
Formula weight	353.48
Temperature/K	180.15
Crystal system	monoclinic
Space group	P2₁/n
a/Å	9.7032(3)
b/Å	9.4092(3)
c/Å	21.5837(7)
α/°	90
β/°	102.899(3)
γ/°	90
Volume/Å³	1920.85(11)
Z	4
ρ_calcg/cm³	1.222
μ/mm⁻¹	0.526
F(000)	760.0
Crystal size/mm³	0.36 × 0.34 × 0.05
Radiation	CuKα (λ = 1.54186)
2Θ range for data collection/°	11.08 to 120.748
Index ranges	-10 ≤ h ≤ 10, -10 ≤ k ≤ 10, -9 ≤ l ≤ 23
Reflections collected	2757
Independent reflections	2757 [R_int = ?, R_sigma = 0.0162]
Data/restraints/parameters	2757/0/353
Goodness-of-fit on F²	1.223
Final R indexes [I>=2σ (I)]	R₁ = 0.0333, wR₂ = 0.0925
Final R indexes [all data]	R₁ = 0.0336, wR₂ = 0.0927
Largest diff. peak/hole / e Å⁻³	0.17/-0.20

9-(4-(4,6-Diphenyl-1,3,5-triazin-2-yl)phenyl)-9N-carbazole (76) ESP496

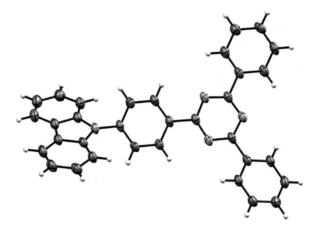

Identification code	ESP496
Empirical formula	$C_{33}H_{22}N_4$
Formula weight	474.54
Temperature/K	160.15
Crystal system	triclinic
Space group	P-1
a/Å	3.8653(2)
b/Å	15.5368(8)
c/Å	19.5050(12)
α/°	101.550(4)
β/°	91.474(5)
γ/°	94.709(4)
Volume/Å³	1142.70(11)
Z	2
ρ_{calc}g/cm³	1.379
μ/mm⁻¹	0.413
F(000)	496.0
Crystal size/mm³	0.22 × 0.04 × 0.03
Radiation	GaKα (λ = 1.34143)
2Θ range for data collection/°	5.07 to 113.934
Index ranges	-1 ≤ h ≤ 4, -19 ≤ k ≤ 19, -24 ≤ l ≤ 24
Reflections collected	13397
Independent reflections	4564 [R_{int} = 0.0509, R_{sigma} = 0.0386]
Indep. refl. with I>=2σ (I)	3112
Data/restraints/parameters	4564/0/334
Goodness-of-fit on F²	0.922
Final R indexes [I>=2σ (I)]	R_1 = 0.0470, wR_2 = 0.1193
Final R indexes [all data]	R_1 = 0.0682, wR_2 = 0.1287
Largest diff. peak/hole / e Å⁻³	0.19/-0.29
CCDC Number	**1871934**

(rac)-1-(N-[2]Paracyclo[2]-6-(N-carbazolyl)(1,4)carbazolophanyl)-4-(4,6-diphenyl-1,3,5-triazin-2-yl)-benzene (117) ESP-CMM18

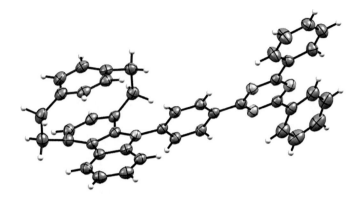

Identification code	ESP-CMM18
Empirical formula	$C_{43}H_{32}N_4$
Formula weight	604.72
Temperature/K	180.15
Crystal system	monoclinic
Space group	$P2_1/c$
a/Å	15.6193(6)
b/Å	13.7087(4)
c/Å	14.6312(5)
α/°	90
β/°	94.394(3)
γ/°	90
Volume/Å^3	3123.63(19)
Z	4
ρ_{calc}g/cm^3	1.286
μ/mm^{-1}	0.586
F(000)	1272.0
Crystal size/mm^3	0.34 × 0.28 × 0.02
Radiation	CuKα (λ = 1.54186)
2Θ range for data collection/°	12.134 to 127.97
Index ranges	-17 ≤ h ≤ 18, -15 ≤ k ≤ 13, -17 ≤ l ≤ 14
Reflections collected	14283
Independent reflections	5057 [R_{int} = 0.0604, R_{sigma} = 0.0492]
Indep. refl. with I>=2σ (I)	3700
Data/restraints/parameters	5057/0/425
Goodness-of-fit on F^2	1.035
Final R indexes [I>=2σ (I)]	R_1 = 0.0547, wR_2 = 0.1415
Final R indexes [all data]	R_1 = 0.0872, wR_2 = 0.1569
Largest diff. peak/hole / e Å^{-3}	0.22/-0.22
CCDC Number	**1871935**

6 List of Abbreviations

(v/v)	volume/volume ratio
(w/w)	weight/weight ratio
°C	degree celsius
ΔE_{ST}	energy gap between S_1 and T_1
δ	chemical shift
μg	microgram
μL	microliter
μmol	micromole
3DMM2O	3D matter made to order (DFG funding program)
Å	Ångström
A	Acceptor
Ac	Acyl
ACN	acetonitrile
Alq_3	tris(8-hydroxyquinolinato)aluminium
APCI	atmospheric-pressure chemical ionization
a.q.	aqueous
Ar	aromat(ic)
ATR	attenuated total reflection
ATRP	atom transfer radical polymerization
a.u.	arbitrary unit
BINOL	1,1'-bi-2-naphthol
Bu	butyl
Bz	Benzoyl
c.w.	clockwise
calc.	calculated

cat.	catalyst
CB8	cucurbit[8]uril
CBP	4,4′-bis-(*N*-carbazolyl)-1,1′-biphenyl
CBS	Corey-Bakshi-Shibata catalyst
c.c.w.	counter clockwise
CCDC	Cambridge Crystallographic Data Centre
cd	candela
CD	circular dichroism
CE_{max}	maximal current efficiency
CIE	*commission internationale de l'éclairage*
CIP	Cahn-Ingold-Prelog
CPL	circularly polarized luminescence
CT	charge transfer
CVD	chemical vapor deposition
Cz	carbazole
Czp	[2]Paracyclo[2](1,4)carbazolophane
d	day
d	doublet
D	Donor
dba	dibenzylideneacetone
DCM	dichloro methane
de	diastereomeric excess (%*de*)
DEPT	distortionless enhancement by polarization transfer
DF	delayed fluorescence
DFT	density functional theory
DIBAL-H	diisobutylaluminium hydride
DMAC	Dimethylacridan

DMF	*N,N*-dimethylformamide
DMSO	dimethyl sulfoxide
DPA	*N,N*-diphenylamino
DPEPO	bis[2-(diphenylphosphino)phenyl] ether oxide
dppf	1,1′-bis(diphenylphosphino)ferrocene
DSC	differential scanning calorimetry
E	Electrophile
e.g.	exempli gratia (for example)
EA	elemental analysis
EBL	electron blocking layer
EDG	electron donating group
ee	enantiomeric excess (*%ee*)
EI	electron ionization
EIL	electon injection layer
EL	electroluminescence
EML	emissive layer
EQE	external quantum efficiency
equiv.	equivalents
Et	ethyl
et al.	*et alii* (and others)
etc.	*et cetera* (and so on)
ETL	electron transport layer
EWG	electron withdrawing group
eV	electron volt
f	oscillator strength
FAB	fast atom bombardment
FDCD	fluorescence-detected circular dichroism

FID	free induction decay
FMO	frontier molecular orbital
g	gram
GC	gas chromatography
GP	general procedure
h	hour
h	Planck constant
HBL	hole blocking layer
HIL	hole injection layer
HOMO	highest occupied molecular orbital
HPLC	high performance liquid chromatography
HRMS	high resolution mass spectrometry
HTL	hole transport layer
HUVEC	human umbilical vein endothelial cell
Hz	Hertz
ICT	intramolecular charge transfer
IDA	indicator displacement assay
IR	infrared
IRRAS	infrared reflection-absorption spectroscopy
ISC	intersystem crossing
IRF	instrument response function
ITO	indium tin oxide
J	coupling constant
K	Kelvin
KSOP	Karlsruhe School of Optics and Photonics
L	liter
LC	liquid crystal

LE	locally excited
LED	light-emitting diode
lm	lumen
log *Ka*	Logarithm of association constant
LUMO	lowest unoccupied molecular orbital
m	*meta*
M	molar
m	multiplet
M.p.	melting point
mbar	millibar
mCP	1,3-bis-(*N*-carbazolyl)benzene
MDA-98	commercial liquid crystal from Merck
mdeg	millidegree
Me	methyl
mg	milligram
MHz	mega Hertz
min	minute
mL	milliliter
mM	milli molar
mol	millimole
*n*BuLi	n-butyllithium
NBS	*N*-bromosuccinimide
NPB	*N,N'*-bis(naphthalen-1-yl)-*N,N'*-bis(phenyl)benzidine
NMR	nuclear magnetic resonance
NOE	nuclear Overhauser effect
NOESY	nuclear Overhauser enhancement and exchange spectroscopy
MS	mass spectrometry

Nu	Nucleophile
OLED	organic light-emitting diode
o	*ortho*
p	*para*
PCP	[2.2]paracyclophane
PE_{max}	maximum power efficiency
PEDOT:PSS	poly(3,4-ethylenedioxythiophene) polystyrene sulfonate
PEGMA	poly(ethylene glycol) methacrylate
PEPPSI	pyridine-enhanced precatalyst preparation stabilization and initiation
PF	prompt fluorescence
Ph	phenyl
pH	decimal logarithm of the reciprocal of the hydrogen ion activity in solution
PIDA	(diacetoxyiodo)benzene
Piv	pivalic
PL	photoluminescence
PLQY	photoluminescence quantum yield
PMMA	poly(methyl methacrylate)
ppm	parts per million
PPT	2,8-bis(diphenylphosphoryl)dibenzo[b,d]thiophene
PPX	poly-(p-xylene)
Py	pyridyl
q	quartet
R/R_P	right-handed (clockwise) stereodescriptor
r.t.	room temperature
rac	racemic
RISC	reverse intersystem crossing
RuPhos	2-dicyclohexylphosphino-2′,6′-diisopropoxybiphenyl

s	seconds
s	singlet
S_1	first excitet singlet state
S/S$_P$	left-handed (counter-clockwise) stereodescriptor
sat.	saturated
SCE	saturated calomel electrode
SEM	Scanning Electron Microscopy
SFB1176	*Sonderforschungsbereich* Project number 1176 (DFG funding program)
SPhos	2-dicyclohexylphosphino-2',6'-dimethoxybiphenyl
t	triplet
T	transmission
T	temperature
T_1	first excited triplet state
TADF	thermally activated delayed fluorescence
^{t}Bu	*tert*-butyl
TCO	transparent conductive oxide
TCTA	tris(4-carbazoyl-9-ylphenyl)amine
TDA	Tamm-Dancoff approximation
TFAA	trifluoracetic anhydride
TGA	thermogravimetric analysis
THF	tetrahydrofuran
TMEDA	tetramethylethylenediamine
TLC	thin layer chromatography
TmPyPB	1,3,5-tris(3-pyridyl-3-phenyl)benzene
tol	toluyl
Trz	1,3,5-triazine
UV	ultraviolet

V	volt
Vis	visible light
V_{on}	onset voltage
vs	very strong
vs	*versus*
vw	very weak
w	weak
W	watt
wt%	weight percent
XPS	x-ray photoelectron spectroscopy

7 Bibliography

[1] C. J. Brown, A. C. Farthing, *Nature* **1949**, *164*, 915.

[2] D. J. Cram, J. M. Cram, *Acc. Chem. Res.* **1971**, *4*, 204–213.

[3] D. K. Lonsdale, H. J. Milledge, K. V. K. Rao, *Proc. R. Soc. London, Ser. A* **1960**, *255*, 82–100.

[4] J. Zyss, I. Ledoux, S. Volkov, V. Chernyak, S. Mukamel, G. P. Bartholomew, G. C. Bazan, *J. Am. Chem. Soc.* **2000**, *122*, 11956–11962.

[5] G. P. Bartholomew, G. C. Bazan, *Acc. Chem. Res.* **2001**, *34*, 30–39.

[6] H. Hopf, *Angew. Chem. Int. Ed.* **2008**, *47*, 9808–9812.

[7] S. E. Gibson, J. D. Knight, *Org. Biomol. Chem.* **2003**, *1*, 1256–1269.

[8] J. Paradies, *Synthesis* **2011**, 3749–3766.

[9] Z. Hassan, E. Spuling, D. M. Knoll, J. Lahann, S. Bräse, *Chem. Soc. Rev.* **2018**, *47*, 6947–6963.

[10] E. J. Corey, *Angew. Chem. Int. Ed. Engl.* **1991**, *30*, 455–465.

[11] Editorial, *Nat. Rev. Drug Discovery* **2004**, *3*, 375.

[12] H. Hopf, *Naturwissenschaften* **1983**, *70*, 349–358.

[13] B. Odell, M. V. Reddington, A. M. Z. Slawin, N. Spencer, J. F. Stoddart, D. J. Williams, *Angew. Chem. Int. Ed. Engl.* **1988**, *27*, 1547–1550.

[14] J. F. Stoddart, *Angew. Chem. Int. Ed.* **2017**, *56*, 11094–11125.

[15] B. S. Joshi, S. W. Pelletier, M. G. Newton, D. Lee, G. B. McGaughey, M. S. Puar, *J. Nat. Prod.* **1996**, *59*, 759–764.

[16] E. Vogel, H. D. Roth, *Angew. Chem. Int. Ed. Engl.* **1964**, *3*, 228–229.

[17] T. Gulder, P. S. Baran, *Nat. Prod. Rep.* **2012**, *29*, 899–934.

[18] A. M. Dilmaç, E. Spuling, A. de Meijere, S. Bräse, *Angew. Chem. Int. Ed.* **2017**, *56*, 5684–5718.

[19] F. Tuinstra, J. L. Koenig, *J. Chem. Phys.* **1970**, *53*, 1126–1130.

[20] F. H. Allen, O. Kennard, D. G. Watson, L. Brammer, A. G. Orpen, R. Taylor, *J. Chem. Soc., Perkin Trans. 2* **1987**, 1–19.

[21] C.-F. Shieh, D. McNally, R. Boyd, *Tetrahedron* **1969**, *25*, 3653–3665.

[22] H. Hope, J. Bernstein, K. N. Trueblood, *Acta Crystallogr., Sect. B: Struct. Crystallogr. Cryst. Chem.* **1972**, *28*, 1733–1743.

[23] F. Vögtle, P. Neumann, *Tetrahedron Lett.* **1969**, *60*, 5329–5334.

[24]   R. S. Cahn, C. Ingold, V. Prelog, *Angew. Chem. Int. Ed.* **1966**, *5*, 385–415.

[25]   M. Enders, Ph.D. Thesis, *Karlsruhe Institute of Technology* **2014**.

[26]   M. Szwarc, *J. Chem. Phys.* **1948**, *16*, 128–136.

[27]   D. J. Cram, H. Steinberg, *J. Am. Chem. Soc.* **1951**, *73*, 5691–5704.

[28]   H. E. Winberg, F. S. Fawcett, W. E. Mochel, C. W. Theobald, *J. Am. Chem. Soc.* **1960**, *82*, 1428–1435.

[29]   H.-F. Chow, K.-H. Low, K. Y. Wong, *Synlett* **2005**, 2130–2134.

[30]   M. Montanari, A. Bugana, A. K. Sharma, D. Pasini, *Org. Biomol. Chem.* **2011**, *9*, 5018–5020.

[31]   T. Otsubo, V. Boekelheide, *Tetrahedron Lett.* **1975**, *16*, 3881–3884.

[32]   F. Bally-Le Gall, C. Hussal, J. Kramer, K. Cheng, R. Kumar, T. Eyster, A. Baek, V. Trouillet, M. Nieger, S. Bräse, J. Lahann, *Chem. Eur. J.* **2017**, *23*, 13342–13350.

[33]   O. R. P. David, *Tetrahedron* **2012**, *68*, 8977–8993.

[34]   H. J. Reich, D. J. Cram, *J. Am. Chem. Soc.* **1969**, *91*, 3527–3533.

[35]   M. Stöbbe, O. Reiser, R. Näder, A. de Meijere, *Chem. Ber.* **1987**, *120*, 1667–1674.

[36]   A. Pelter, B. Mootoo, A. Maxwell, A. Reid, *Tetrahedron Lett.* **2001**, *42*, 8391–8394.

[37]   P. Lennartz, G. Raabe, C. Bolm, *Adv. Synth. Catal.* **2012**, *354*, 3237–3249.

[38]   J. J. P. Kramer, C. Yildiz, M. Nieger, S. Bräse, *Eur. J. Org. Chem.* **2014**, 1287–1295.

[39]   P. Lennartz, G. Raabe, C. Bolm, *Isr. J. Chem.* **2012**, *52*, 171–179.

[40]   P. Schaal, H. Baars, G. Raabe, I. Atodiresei, C. Bolm, *Adv. Synth. Catal.* **2013**, *355*, 2506–2512.

[41]   D. J. Cram, A. C. Day, *J. Org. Chem.* **1966**, *31*, 1227–1232.

[42]   N. V. Vorontsova, V. I. Rozenberg, E. V. Vorontsov, D. Y. Antonov, Z. A. Starikova, Y. N. Bubnov, *Russ. Chem. Bull.* **2002**, *51*, 1483–1490.

[43]   M. Cakici, Z. G. Gu, M. Nieger, J. Burck, L. Heinke, S. Bräse, *Chem. Commun.* **2015**, *51*, 4796–4798.

[44]   C. Braun, E. Spuling, N. B. Heine, M. Cakici, M. Nieger, S. Bräse, *Adv. Synth. Catal.* **2016**, *358*, 1664–1670.

[45]   B. Wang, J. W. Graskemper, L. Qin, S. G. DiMagno, *Angew. Chem. Int. Ed.* **2010**, *49*, 4079–4083.

[46]   S. Banfi, A. Manfredi, F. Montanari, G. Pozzi, S. Quici, *J. Mol. Catal. A: Chem.* **1996**, *113*, 77–86.

[47] D. Y. Antonov, Y. N. Belokon, N. S. Ikonnikov, S. A. Orlova, A. P. Pisarevsky, N. I. Raevski, V. I. Rozenberg, E. V. Sergeeva, Y. T. Struchkov, V. I. Tararov, E. V. Vorontsov, *J. Chem. Soc., Perkin Trans. 1* **1995**, 1873–1879.

[48] C. J. Friedmann, S. Ay, S. Bräse, *J. Org. Chem.* **2010**, *75*, 4612–4614.

[49] P. B. Hitchcock, G. J. Rowlands, R. J. Seacome, *Org. Biomol. Chem.* **2005**, *3*, 3873–3876.

[50] P. B. Hitchcock, G. J. Rowlands, R. Parmar, *Chem. Commun.* **2005**, 4219–4221.

[51] G. J. Rowlands, *Org. Biomol. Chem.* **2008**, *6*, 1527–1534.

[52] G. J. Rowlands, R. J. Seacome, *Beilstein J. Org. Chem.* **2009**, *5*, 9.

[53] R. Parmar, M. P. Coles, P. B. Hitchcock, G. J. Rowlands, *Synthesis* **2010**, 4177–4187.

[54] B. Ortner, R. Waibel, P. Gmeiner, *Angew. Chem. Int. Ed.* **2001**, *40*, 1283–1285.

[55] S. V. Luis, M. I. Burguete, *Tetrahedron* **1991**, *47*, 1737–1744.

[56] N. De Rycke, J. Marrot, F. Couty, O. R. P. David, *Eur. J. Org. Chem.* **2011**, 1980–1984.

[57] W. F. Gorham, *J. Polym. Sci., Part A-1: Polym. Chem.* **1966**, *4*, 3027–3039.

[58] J. Lahann, R. Langer, *Macromolecules* **2002**, *35*, 4380–4386.

[59] Y. Elkasabi, J. Lahann, *Macromol. Rapid Commun.* **2009**, *30*, 57–63.

[60] J. Bienkiewicz, *Med. Device Technol.* **2006**, *17*, 10–11.

[61] A. K. Bier, M. Bognitzki, J. Mogk, A. Greiner, *Macromolecules* **2012**, *45*, 1151–1157.

[62] J. Jakabovič, J. Kováč, M. Weis, D. Haško, R. Srnánek, P. Valent, R. Resel, *Microelectron. J.* **2009**, *40*, 595–597.

[63] T. Y. Chang, V. G. Yadav, S. De Leo, A. Mohedas, B. Rajalingam, C.-L. Chen, S. Selvarasah, M. R. Dokmeci, A. Khademhosseini, *Langmuir* **2007**, *23*, 11718–11725.

[64] J. Hsu, L. Rieth, R. A. Normann, P. Tathireddy, F. Solzbacher, *IEEE Trans. Biomed. Eng.* **2009**, *56*, 23–29.

[65] A. L. Winkler, M. Koenig, A. Welle, V. Trouillet, D. Kratzer, C. Hussal, J. Lahann, C. Lee-Thedieck, *Biomacromolecules* **2017**, *18*, 3089–3098.

[66] M. Koenig, R. Kumar, C. Hussal, V. Trouillet, L. Barner, J. Lahann, *Macromol. Chem. Phys.* **2017**, *218*.

[67] F. Xie, X. P. Deng, D. Kratzer, K. C. K. Cheng, C. Friedmann, S. H. Qi, L. Solorio, J. Lahann, *Angew. Chem. Int. Ed.* **2017**, *56*, 203–207.

[68] X. W. Jiang, H. Y. Chen, G. Galvan, M. Yoshida, J. Lahann, *Adv. Funct. Mater.* **2008**, *18*, 27–35.

[69]   K. C. Cheng, M. A. Bedolla-Pantoja, Y.-K. Kim, J. V. Gregory, F. Xie, A. de France, C. Hussal, K. Sun, N. L. Abbott, J. Lahann, *Science* **2018**, *362*, 804–808.

[70]   A. Bernanose, *Br. J. Appl. Phys.* **1955**, *6*, 54–56.

[71]   M. Pope, H. P. Kallmann, P. Magnante, *J. Chem. Phys.* **1963**, *38*, 2042–2043.

[72]   M. Sano, M. Pope, H. Kallmann, *J. Chem. Phys.* **1965**, *43*, 2920–2921.

[73]   C. W. Tang, S. A. VanSlyke, *Appl. Phys. Lett.* **1987**, *51*, 913–915.

[74]   A. J. Heeger, *Angew. Chem. Int. Ed.* **2001**, *113*, 2660–2682.

[75]   D. Volz, Ph.D. Thesis, *Karlsruhe Institute of Technology* **2014**.

[76]   R. A. Marcus, *Rev. Mod. Phys.* **1993**, *65*, 599–610.

[77]   A. Troisi, D. L. Cheung, D. Andrienko, *Phys. Rev. Lett.* **2009**, *102*, 116602.

[78]   D. Hertel, C. D. Müller, K. Meerholz, *Chem. Unserer Zeit* **2005**, *39*, 336–347.

[79]   K. Müllen, U. Scherf, *Organic Light-Emitting Devices: Synthesis, Properties and Applications*, Wiley-VCH, Weinheim, **2006**.

[80]   L. H. Smith, J. A. Wasey, I. D. Samuel, W. L. Barnes, *Adv. Funct. Mater.* **2005**, *15*, 1839–1844.

[81]   T. J. Marks, J. G. C. Veinot, J. Cui, H. Yan, A. Wang, N. L. Edleman, J. Ni, Q. Huang, P. Lee, N. R. Armstrong, *Synth. Met.* **2002**, *127*, 29–35.

[82]   J. Saghaei, A. Fallahzadeh, T. Saghaei, *Org. Electron.* **2015**, *24*, 188–194.

[83]   R. Grover, R. Srivastava, M. N. Kamalasanan, D. S. Mehta, *J. Lumin.* **2014**, *146*, 53–56.

[84]   A. C. Arias, J. D. MacKenzie, I. McCulloch, J. Rivnay, A. Salleo, *Chem. Rev.* **2010**, *110*, 3–24.

[85]   Y. Yang, *Progress in High-Efficient Solution Process Organic Photovoltaic Devices: Fundamentals, Materials, Devices and Fabrication*, Springer, Berlin, **2015**.

[86]   A. Chutinan, K. Ishihara, T. Asano, M. Fujita, S. Noda, *Org. Electron.* **2005**, *6*, 3–9.

[87]   W. H. Koo, S. M. Jeong, F. Araoka, K. Ishikawa, S. Nishimura, T. Toyooka, H. Takezoe, *Nat. Photonics* **2010**, *4*, 222–226.

[88]   Y. Tao, K. Yuan, T. Chen, P. Xu, H. Li, R. Chen, C. Zheng, L. Zhang, W. Huang, *Adv. Mater.* **2014**, *26*, 7931–7958.

[89]   C. Adachi, M. A. Baldo, M. E. Thompson, S. R. Forrest, *J. Appl. Phys.* **2001**, *90*, 5048–5051.

[90]   H. Yersin, *Highly Efficient OLEDs with Phosphorescent Materials*, Wiley-VCH, Weinheim, **2008**.

[91]   A. Endo, M. Ogasawara, A. Takahashi, D. Yokoyama, Y. Kato, C. Adachi, *Adv. Mater.* **2009**, *21*, 4802–4806.

[92]   A. Endo, K. Sato, K. Yoshimura, T. Kai, A. Kawada, H. Miyazaki, C. Adachi, *Appl. Phys. Lett.* **2011**, *98*, 42.

[93]   H. Uoyama, K. Goushi, K. Shizu, H. Nomura, C. Adachi, *Nature* **2012**, *492*, 234–238.

[94]   Q. Zhang, J. Li, K. Shizu, S. Huang, S. Hirata, H. Miyazaki, C. Adachi, *J. Am. Chem. Soc.* **2012**, *134*, 14706–14709.

[95]   M. Y. Wong, E. Zysman-Colman, *Adv. Mater.* **2017**, *29*, 1605444.

[96]   C. A. Parker, C. G. Hatchard, *Trans. Faraday Soc.* **1961**, *57*, 1894–1904.

[97]   J. Saltiel, H. C. Curtis, L. Metts, J. W. Miley, J. Winterle, M. Wrighton, *J. Am. Chem. Soc.* **1970**, *92*, 410–411.

[98]   R. E. Brown, L. A. Singer, J. H. Parks, *Chem. Phys. Lett.* **1972**, *14*, 193–195.

[99]   H. Yersin, H. Otto, G. Gliemann, *Theoret. Chim. Acta* **1974**, *33*, 63–78.

[100] A. Maciejewski, M. Szymanski, R. P. Steer, *J. Phys. Chem.* **1986**, *90*, 6314–6318.

[101] Z. Yang, Z. Mao, Z. Xie, Y. Zhang, S. Liu, J. Zhao, J. Xu, Z. Chi, M. P. Aldred, *Chem. Soc. Rev.* **2017**, *46*, 915–1016.

[102] K. Goushi, K. Yoshida, K. Sato, C. Adachi, *Nat. Photonics* **2012**, *6*, 253–258.

[103] Y. Im, M. Kim, Y. J. Cho, J. A. Seo, K. S. Yook, J. Y. Lee, *Chem. Mater.* **2017**, *29*, 1946–1963.

[104] T. Hatakeyama, K. Shiren, K. Nakajima, S. Nomura, S. Nakatsuka, K. Kinoshita, J. Ni, Y. Ono, T. Ikuta, *Adv. Mater.* **2016**, *28*, 2777–2781.

[105] K. Cheng, Ph.D. Thesis, *University of Michigan* **2017**.

[106] T. J. Penfold, F. B. Dias, A. P. Monkman, *Chem. Commun.* **2018**, *54*, 3926–3935.

[107] K. Kawasumi, T. Wu, T. Zhu, H. S. Chae, T. Van Voorhis, M. A. Baldo, T. M. Swager, *J. Am. Chem. Soc.* **2015**, *137*, 11908–11911.

[108] H. Tsujimoto, D. G. Ha, G. Markopoulos, H. S. Chae, M. A. Baldo, T. M. Swager, *J. Am. Chem. Soc.* **2017**, *139*, 4894–4900.

[109] E. Spuling, N. Sharma, I. D. W. Samuel, E. Zysman-Colman, S. Bräse, *Chem. Commun.* **2018**, *54*, 9278–9281.

[110] M.-Y. Zhang, Z.-Y. Li, B. Lu, Y. Wang, Y.-D. Ma, C.-H. Zhao, *Org. Lett.* **2018**, *20*, 6868–6871.

[111] M. Kim, S. K. Jeon, S.-H. Hwang, S.-S. Lee, E. Yu, J. Y. Lee, *J. Phys. Chem. C* **2016**, *120*, 2485–2493.

[112] H. Tanaka, K. Shizu, H. Miyazaki, C. Adachi, *Chem. Commun.* **2012**, *48*, 11392–11394.

[113]   W.-L. Tsai, M.-H. Huang, W.-K. Lee, Y.-J. Hsu, K.-C. Pan, Y.-H. Huang, H.-C. Ting, M. Sarma, Y.-Y. Ho, H.-C. Hu, C.-C. Chen, M.-T. Lee, K.-T. Wong, C.-C. Wu, *Chem. Commun.* **2015**, *51*, 13662–13665.

[114]   L. S. Cui, H. Nomura, Y. Geng, J. U. Kim, H. Nakanotani, C. Adachi, *Angew. Chem. Int. Ed.* **2017**, *56*, 1571–1575.

[115]   W. Huang, M. Einzinger, T. Zhu, H. S. Chae, S. Jeon, S.-G. Ihn, M. Sim, S. Kim, M. Su, G. Teverovskiy, T. Wu, T. Van Voorhis, T. M. Swager, M. A. Baldo, S. L. Buchwald, *Chem. Mater.* **2018**, *30*, 1462–1466.

[116]   P. J. Pye, K. Rossen, R. A. Reamer, N. N. Tsou, R. Volante, P. J. Reider, *J. Am. Chem. Soc.* **1997**, *119*, 6207–6208.

[117]   J. W. Hong, H. Y. Woo, B. Liu, G. C. Bazan, *J. Am. Chem. Soc.* **2005**, *127*, 7435–7443.

[118]   Y. Morisaki, Y. Chujo, *Polym. Chem.* **2011**, *2*, 1249–1257.

[119]   Y. Morisaki, Y. Chujo, *Chem. Lett.* **2012**, *41*, 840–846.

[120]   Y. Morisaki, M. Gon, T. Sasamori, N. Tokitoh, Y. Chujo, *J. Am. Chem. Soc.* **2014**, *136*, 3350–3353.

[121]   M. Gon, Y. Morisaki, R. Sawada, Y. Chujo, *Chem. Eur. J.* **2016**, *22*, 2291–2298.

[122]   M. Gon, Y. Morisaki, Y. Chujo, *Chem. Commun.* **2017**, *53*, 8304–8307.

[123]   M. Gon, R. Sawada, Y. Morisaki, Y. Chujo, *Macromolecules* **2017**, *50*, 1790–1802.

[124]   K.-J. Yoon, H.-J. Noh, D.-W. Yoon, I.-A. Shin, J.-Y. Kim, *EP3029753B1* **2017**.

[125]   S. L. Buchwald, W. Huang, *US9972791B2* **2018**.

[126]   E. Spuling, M.Sc. Thesis, *Karlsruhe Institute of Technology* **2015**.

[127]   D. M. Knoll, S. Bräse, *ACS Omega* **2018**, *3*, 12158–12162.

[128]   M. Y. Wong, S. Krotkus, G. Copley, W. Li, C. Murawski, D. Hall, G. J. Hedley, M. Jaricot, D. B. Cordes, A. M. Z. Slawin, Y. Olivier, D. Beljonne, L. Muccioli, M. Moral, J.-C. Sancho-Garcia, M. C. Gather, I. D. W. Samuel, E. Zysman-Colman, *ACS Appl. Mater. Interfaces* **2018**, *10*, 33360–33372.

[129]   Z. Ahmadi, J. S. McIndoe, *Chem. Commun.* **2013**, *49*, 11488–11490.

[130]   M. J. Frisch, G. W. Trucks, H. B. Schlegel, G. E. Scuseria, M. A. Robb, J. R. Cheeseman, G. Scalmani, e. al., Gaussian Inc., Wallingford, CT, **2013**.

[131]   C. Adamo, V. Barone, *J. Chem. Phys.* **1999**, *110*, 6158–6170.

[132]   J. A. Pople, J. S. Binkley, R. Seeger, *Int. J. Quant. Chem. Symp.* **1976**, *10*, 1.

[133]   M. Moral, L. Muccioli, W. J. Son, Y. Olivier, J. C. Sancho-García, *J. Chem. Theory Comput.*
        2015, *11*, 168–177.

[134]   N. Sharma, Ph.D. Thesis, *University of St. Andrews* **2019**.

[135]   E. Elacqua, L. R. MacGillivray, *Eur. J. Org. Chem.* **2010**, 6883–6894.

[136]   J. L. Zafra, A. Molina Ontoria, P. Mayorga Burrezo, M. Pena-Alvarez, M. Samoc, J. Szeremeta,
        F. J. Ramirez, M. D. Lovander, C. J. Droske, T. M. Pappenfus, L. Echegoyen, J. T. Lopez
        Navarrete, N. Martin, J. Casado, *J. Am. Chem. Soc.* **2017**, *139*, 3095–3105.

[137]   W. H. Melhuish, *J. Phys. Chem.* **1961**, *65*, 229–235.

[138]   H. Allgeier, M. G. Siegel, R. C. Helgeson, E. Schmidt, D. J. Cram, *J. Am. Chem. Soc.* **1975**, *97*,
        3782–3789.

[139]   A. von Orlikowski, B.Sc. Thesis, *Karlsruhe Institute of Technology* **2017**.

[140]   K. P. Jayasundera, D. N. Kusmus, L. Deuilhe, L. Etheridge, Z. Farrow, D. J. Lun, G. Kaur, G.
        J. Rowlands, *Org. Biomol. Chem.* **2016**, *14*, 10848–10860.

[141]   H. J. Reich, D. J. Cram, *J. Am. Chem. Soc.* **1969**, *91*, 3534–3543.

[142]   M. Enders, Ph.D. thesis Thesis, **2014**.

[143]   D. Webb, T. F. Jamison, *Org. Lett.* **2012**, *14*, 568–571.

[144]   C. M. Mattern, B.Sc. Thesis, *Karlsruhe Institute of Technology* **2018**.

[145]   D. R. Lee, J. M. Choi, C. W. Lee, J. Y. Lee, *ACS Appl. Mater. Interfaces* **2016**, *8*, 23190–23196.

[146]   S.-H. Kim, M.-K. Kim, C.-W. Chu, K.-H. Lee, *US9324953B2* **2016**.

[147]   M. Kreis, C. J. Friedmann, S. Bräse, *Chem. Eur. J.* **2005**, *11*, 7387–7394.

[148]   F. G. Bordwell, J. C. Branca, D. L. Hughes, W. N. Olmstead, *J. Org. Chem.* **1980**, *45*, 3305–
        3313.

[149]   F. G. Bordwell, G. E. Drucker, H. E. Fried, *J. Org. Chem.* **1981**, *46*, 632–635.

[150]   F. G. Bordwell, *Acc. Chem. Res.* **1988**, *21*, 456–463.

[151]   M. A. Rodriguez, S. D. Bunge, *Acta Crystallogr., Sect. E: Struct. Rep. Online* **2003**, *59*, o1123–
        o1125.

[152]   S. Y. Byeon, J. Kim, D. R. Lee, S. H. Han, S. R. Forrest, J. Y. Lee, *Adv. Optical Mater.* **2018**,
        6.

[153]   N. G. Connelly, W. E. Geiger, *Chem. Rev.* **1996**, *96*, 877–910.

[154]   C. M. Cardona, W. Li, A. E. Kaifer, D. Stockdale, G. C. Bazan, *Adv. Mater.* **2011**, *23*, 2367–
        2371.

[155]   C. Han, Y. Zhao, H. Xu, J. Chen, Z. Deng, D. Ma, Q. Li, P. Yan, *Chem. Eur. J.* **2011**, *17*, 5800–5803.

[156]   S.-J. Su, T. Chiba, T. Takeda, J. Kido, *Adv. Mater.* **2008**, *20*, 2125–2130.

[157]   C. Braun, Ph.D. Thesis, *Karlsruhe Institute of Technology* **2017**.

[158]   M. Toda, Y. Inoue, T. Mori, *ACS Omega* **2018**, *3*, 22–29.

[159]   C. Rosini, R. Ruzziconi, S. Superchi, F. Fringuelli, O. Piermatti, *Tetrahedron: Asymmetry* **1998**, *9*, 55–62.

[160]   S. Feuillastre, M. Pauton, L. Gao, A. Desmarchelier, A. J. Riives, D. Prim, D. Tondelier, B. Geffroy, G. Muller, G. Clavier, G. Pieters, *J. Am. Chem. Soc.* **2016**, *138*, 3990–3993.

[161]   M. Li, S. H. Li, D. Zhang, M. Cai, L. Duan, M. K. Fung, C. F. Chen, *Angew. Chem. Int. Ed.* **2018**, *57*, 2889–2893.

[162]   E. D. Laganis, R. G. Finke, V. Boekelheide, *Tetrahedron Lett.* **1980**, *21*, 4405–4408.

[163]   P. J. Dyson, B. F. G. Johnson, C. M. Martin, *Coord. Chem. Rev.* **1998**, *175*, 59–89.

[164]   T. Murahashi, M. Fujimoto, Y. Kawabata, R. Inoue, S. Ogoshi, H. Kurosawa, *Angew. Chem. Int. Ed.* **2007**, *119*, 5536–5539.

[165]   G. A. Papoyan, K. P. Butin, R. Hoffmann, V. I. Rozenberg, *Russ. Chem. Bull.* **1998**, *47*, 153–159.

[166]   M. Cakici, S. Bräse, *Eur. J. Org. Chem.* **2012**, 6132–6135.

[167]   S. Itsuno, K. Ito, A. Hirao, S. Nakahama, *J. Chem. Soc., Chem. Commun.* **1983**, 469–470.

[168]   S. Itsuno, K. Ito, A. Hirao, S. Nakahama, *J. Chem. Soc., Perkin Trans. 1* **1984**, 2887–2893.

[169]   S. Itsuno, M. Nakano, K. Miyazaki, H. Masuda, K. Ito, A. Hirao, S. Nakahama, *J. Chem. Soc., Perkin Trans. 1* **1985**, 2039–2044.

[170]   E. J. Corey, R. K. Bakshi, S. Shibata, *J. Am. Chem. Soc.* **1987**, *109*, 5551–5553.

[171]   E. J. Corey, C. J. Helal, *Angew. Chem. Int. Ed.* **1998**, *37*, 1986–2012.

[172]   P. Dorizon, C. Martin, J.-C. Daran, J.-C. Fiaud, H. B. Kagan, *Tetrahedron: Asymmetry* **2001**, *12*, 2625–2630.

[173]   S. M. Nettles, K. Matos, E. R. Burkhardt, D. R. Rouda, J. A. Corella, *J. Org. Chem.* **2002**, *67*, 2970–2976.

[174]   P. G. Jones, I. Dix, M. Negru, D. Schollmeyer, *Z. Naturforsch. B* **2011**, *66*, 705–710.

[175]   R. C. Breton, W. F. Reynolds, *Nat. Prod. Rep.* **2013**, *30*, 501–524.

[176]   M. M. J. Smulders, A. P. H. J. Schenning, E. W. Meijer, *J. Am. Chem. Soc.* **2008**, *130*, 606–611.

[177]   L. Brunsveld, J. Vekemans, J. Hirschberg, R. Sijbesma, E. Meijer, *Proc. Natl. Acad. Sci.* **2002**, *99*, 4977–4982.

[178]   J. H. K. K. Hirschberg, L. Brunsveld, A. Ramzi, J. A. J. M. Vekemans, R. P. Sijbesma, E. W. Meijer, *Nature* **2000**, *407*, 167.

[179]   P. A. Korevaar, S. J. George, A. J. Markvoort, M. M. J. Smulders, P. A. J. Hilbers, A. P. H. J. Schenning, T. F. A. De Greef, E. W. Meijer, *Nature* **2012**, *481*, 492.

[180]   K. Hanabusa, M. Yamada, M. Kimura, H. Shirai, *Angew. Chem. Int. Ed. Engl.* **1996**, *35*, 1949–1951.

[181]   D. Frank, M. Nieger, C. Friedmann, J. Lahann , S. Bräse, *Isr. J. Chem.* **2012**, *52*, 143–148.

[182]   B. E. Smart, *J. Fluorine Chem.* **2001**, *109*, 3–11.

[183]   Y. L. Yeh, W. F. Gorham, *J. Org. Chem.* **1969**, *34*, 2366–2370.

[184]   L. A. Singer, D. J. Cram, *J. Am. Chem. Soc.* **1963**, *85*, 1080–1084.

[185]   A. R. Katritzky, B. Pilarski, L. Urogdi, *Synthesis* **1989**, 949–950.

[186]   S. Sinn, F. Biedermann, *Isr. J. Chem.* **2018**, *58*, 357–412.

[187]   H. H. Jo, C.-Y. Lin, E. V. Anslyn, *Acc. Chem. Res.* **2014**, *47*, 2212–2221.

[188]   B. T. Nguyen, E. V. Anslyn, *Coord. Chem. Rev.* **2006**, *250*, 3118–3127.

[189]   B. Wang, E. V. Anslyn, *Chemosensors: Principles, Strategies, and Applications, Vol. 15*, Wiley, Hoboken, **2011**.

[190]   L. You, D. Zha, E. V. Anslyn, *Chem. Rev.* **2015**, *115*, 7840–7892.

[191]   R. N. Dsouza, A. Hennig, W. M. Nau, *Chem. Eur. J.* **2012**, *18*, 3444–3459.

[192]   V. M. Mirsky, A. Yatsimirsky., *Artificial Receptors for Chemical Sensors*, Wiley-VCH, Weinheim, **2011**.

[193]   T. Schrader, S. Koch, *Mol. BioSyst.* **2007**, *3*, 241–248.

[194]   N. Rifai, A. R. Horvath, C. T. Wittwer, *Tietz Fundamentals of Clinical Chemistry and Molecular Diagnostics*, 6. ed., Elsevier, Missouri, **2018**.

[195]   H.-J. Schneider, A. K. Yatsimirsky, *Chem. Soc. Rev.* **2008**, *37*, 263–277.

[196]   H.-J. Schneider, P. Agrawal, A. K. Yatsimirsky, *Chem. Soc. Rev.* **2013**, *42*, 6777–6800.

[197]   S. Rochat, J. Gao, X. Qian, F. Zaubitzer, K. Severin, *Chem. Eur. J.* **2010**, *16*, 104–113.

[198]   J. Li, D. Yim, W.-D. Jang, J. Yoon, *Chem. Soc. Rev.* **2017**, *46*, 2437–2458.

[199] H. He, M. A. Mortellaro, M. J. P. Leiner, R. J. Fraatz, J. K. Tusa, *J. Am. Chem. Soc.* **2003**, *125*, 1468–1469.

[200] A. P. de Silva, H. Q. N. Gunaratne, T. Gunnlaugsson, A. J. M. Huxley, C. P. McCoy, J. T. Rademacher, T. E. Rice, *Chem. Rev.* **1997**, *97*, 1515–1566.

[201] T. Zhang, E. V. Anslyn, *Org. Lett.* **2007**, *9*, 1627–1629.

[202] G. V. Zyryanov, M. A. Palacios, P. Anzenbacher, *Angew. Chem. Int. Ed.* **2007**, *46*, 7849–7852.

[203] J. Kornhuber, G. Quack, *Neurosci. Lett.* **1995**, *195*, 137–139.

[204] C. G. Parsons, G. Rammes, W. Danysz, *Curr. Neuropharmacol.* **2008**, *6*, 55–78.

[205] J. Kim, I. S. Jung, S. Y. Kim, E. Lee, J. K. Kang, S. Sakamoto, K. Yamaguchi, K. Kim, *J. Am. Chem. Soc.* **2000**, *122*, 540–541.

[206] J. Murray, K. Kim, T. Ogoshi, W. Yao, B. C. Gibb, *Chem. Soc. Rev.* **2017**, *46*, 2479–2496.

[207] S. J. Barrow, S. Kasera, M. J. Rowland, J. del Barrio, O. A. Scherman, *Chem. Rev.* **2015**, *115*, 12320–12406.

[208] J. Lagona, P. Mukhopadhyay, S. Chakrabarti, L. Isaacs, *Angew. Chem. Int. Ed.* **2005**, *44*, 4844–4870.

[209] E. Masson, X. Ling, R. Joseph, L. Kyeremeh-Mensah, X. Lu, *RSC Adv.* **2012**, *2*, 1213–1247.

[210] K. I. Assaf, W. M. Nau, *Chem. Soc. Rev.* **2015**, *44*, 394–418.

[211] F. Biedermann, D. Hathazi, W. M. Nau, *Chem. Commun.* **2015**, *51*, 4977–4980.

[212] F. Biedermann, W. M. Nau, *Angew. Chem. Int. Ed.* **2014**, *53*, 5694–5699.

[213] J. M. Chinai, A. B. Taylor, L. M. Ryno, N. D. Hargreaves, C. A. Morris, P. J. Hart, A. R. Urbach, *J. Am. Chem. Soc.* **2011**, *133*, 8810–8813.

[214] G. Ghale, V. Ramalingam, A. R. Urbach, W. M. Nau, *J. Am. Chem. Soc.* **2011**, *133*, 7528–7535.

[215] P. Rajgariah, A. Urbach, *J. Inclusion Phenom. Macrocyclic Chem.* **2008**, *62*, 251–254.

[216] J. Li, D. J. Mooney, *Nat. Rev. Mater.* **2016**, *1*, 16071.

[217] A. I. Lazar, F. Biedermann, K. R. Mustafina, K. I. Assaf, A. Hennig, W. M. Nau, *J. Am. Chem. Soc.* **2016**, *138*, 13022–13029.

[218] M. M. Ayhan, H. Karoui, M. Hardy, A. Rockenbauer, L. Charles, R. Rosas, K. Udachin, P. Tordo, D. Bardelang, O. Ouari, *J. Am. Chem. Soc.* **2015**, *137*, 10238–10245.

[219] J. W. Lee, S. Samal, N. Selvapalam, H.-J. Kim, K. Kim, *Acc. Chem. Res.* **2003**, *36*, 621–630.

[220] S. Sinn, E. Spuling, S. Bräse, F. Biedermann, *in preparation* **2019**.

[221] A. Izuoka, S. Murata, T. Sugawara, H. Iwamura, *J. Am. Chem. Soc.* **1987**, *109*, 2631–2639.

[222]   L. Ernst, *Liebigs Ann.* **1995**, *1995*, 13–17.

[223]   D. J. Cram, N. L. Allinger, *J. Am. Chem. Soc.* **1955**, *77*, 6289–6294.

[224]   E. Benedetti, M. L. Delcourt, B. Gatin-Fraudet, S. Turcaud, L. Micouin, *RSC Adv.* **2017**, *7*, 50472–50476.

[225]   C. Braun, S. Bräse, L. L. Schafer, *Eur. J. Org. Chem.* **2017**, 1760–1764.

[226]   W. Guo, *Org. Biomol. Chem.* **2015**, *13*, 10285–10289.

[227]   J. Xiao, X. K. Liu, X. X. Wang, C. J. Zheng, F. Li, *Org. Electron.* **2014**, *15*, 2763–2768.

[228]   L. Shi, G. Tian, H. Ye, S. Qi, D. Wu, *Polymer* **2014**, *55*, 1150–1159.

[229]   Y. Liu, D. Chao, H. Yao, *Org. Electron.* **2014**, *15*, 1422–1431.

[230]   J. Lahann, D. Klee, H. Höcker, *Macromol. Rapid Commun.* **1998**, *19*, 441–444.

[231]   H. Matsuo, H. Kobayashi, T. Tatsuno, *Chem. Pharm. Bull.* **1970**, *18*, 1693–1695.

[232]   M. Busch, Ph.D. Thesis, *Karlsruhe Institute of Technology* **2014**.

[233]   T. Focken, H. Hopf, V. Snieckus, I. Dix, Peter G. Jones, *Eur. J. Org. Chem.* **2001**, 2221–2228.

[234]   A. A. Aly, A. A. Hassan, Y. S. Mohamed, A.-F. E. Mourad, H. Hopf, *Monsh. Chem.* **1992**, *123*, 179–189.

8 Appendix

8.1 Curriculum Vitae

Personal Information

Name	Eduard Spuling
Born	19th May 1989 in Pavlodar

Education

01/2016 – 05/2019 Doctorate at the Institute of Organic Chemistry of the Karlsruhe Institute of Technology (KIT) under the supervision of Prof. Dr. Stefan Bräse (1.0)

10/2018 – 11/2018 Visiting researcher at the University of St. Andrews under the supervision of Dr. Eli Zysman-Colman

01/2016 – 01/2018 MBA Fundamentals Fellow at the Hector School of Engineering and Management of the Karlsruhe Institute of Technology (KIT) (1.6)

04/2013 – 10/2015 Master of Science in Chemistry from the Karlsruhe Institute of Technology (KIT) (1.0)

Master thesis (Prof. Dr. Stefan Bräse) on Organic Synthesis of OLED Emitters (1.0)

08/2014 – 12/2014 Research Intern at the BASF Global Research Centre Singapore

Synthesis of OFET Polymers (Dr. Fulvio G. Brunetti)

10/2009 – 04/2013 Bachelor of Science in Chemistry from the Karlsruhe Institute of Technology (KIT) (1.6)

Bachelor thesis (Prof. Dr. Hans-Achim Wagenknecht) on Synthesis of DNA building blocks for post synthetic modification (1.0)

09/2000 – 07/2009 Abitur from Einsteingymnasium Kehl am Rhein (1.9)

Major fields of study: Physics and Chemistry

Awards

06/2018 KIT Innovation Contest "Neuland" (2nd Place)

08/2017 – 12/2018 PhD Scholarship by Karlsruhe School of Optics and Photonics (KSOP)

12/2016 Rétey Award for Outstanding Study Achievements in Organic Chemistry

8.2 List of Publications

Publications

1) Z. Hassan, E. Spuling, S. Bräse, *Angew. Chem. Int. Ed.* **2019**, *58*, 2–17.

   Regioselective Functionalization of [2.2]Paracyclophanes: Recent Synthetic Advances and Perspectives

2) S. Sinn, E. Spuling, S. Bräse, F. Biedermann, *Chem. Sci.* **2019**, *10*, 6584–6593.

   Rational Design and Implementation of a Cucurbit[8]uril-based Indicator-Displacement Assay for Application in Blood Serum

3) N. Sharma, E. Spuling, C. M. Mattern, W. Li, O. Fuhr, S. Bräse, I. D. W. Samuel, E. Zysman-Colman, *Chem. Sci.* **2019**, *10*, 6689–6696.

   Turn on of Sky-Blue Thermally Activated Delayed Fluorescence via Increased Torsion by a Bulky Carbazolophane Donor

4) E. Spuling, N. Sharma, I. D. W. Samuel, E. Zysman-Colman, S. Bräse, *Chem. Commun.* **2018**, *54*, 9278–9281.

   (Deep) blue through-space conjugated TADF emitters based on [2.2]paracyclophanes

5) Z. Hassan, E. Spuling, D. M. Knoll, J. Lahann, S. Bräse, *Chem. Soc. Rev.* **2018**, *47*, 6947–6963.

   Planar chiral [2.2]paracyclophanes: from synthetic curiosity to applications in asymmetric synthesis and materials

6) C. Bizzarri, E. Spuling, D. M. Knoll, D. Volz, S. Bräse, *Coord. Chem. Rev.* **2017**, *373*, 49–82.

   Sustainable metal complexes for organic light-emitting diodes (OLEDs)

7) A. M. Dilmaç, E. Spuling, A. de Meijere, S. Bräse, *Angew. Chem. Int. Ed.* **2017**, *56*, 5684–5718.

   Propellanes—From a Chemical Curiosity to "Explosive" Materials and Natural Products

8) A. A. Aly, A. A. Hassan, S. Bräse, M. A. A. Ibrahim, E.-S. S. M. Abd Al-Latif, E. Spuling, M. Nieger, *J. Sulf. Chem.* **2016**, 1–7.

   1,3,4-Thiadiazoles and 1,3-thiazoles from one-pot reaction of bisthioureas with 2-(bis(methylthio)methylene)malononitrile and ethyl 2-cyano-3,3-bis(methylthio)acrylate

9) C. Braun, E. Spuling, N. B. Heine, M. Cakici, M. Nieger, S. Bräse, *Adv. Synth. Catal.* **2016**, *358*, 1664–1670.

   Efficient Modular Synthesis of Isomeric Mono- and Bispyridyl[2.2]paracyclophanes by Palladium-Catalyzed Cross- Coupling Reactions

Conference Posters

1) MAKRO 09/2018 in Karlsruhe

   Sequence-control in Three Dimensions – A Building Block Approach

2) Karlsruhe Days of Optics and Photonics 11/2017 in Karlsruhe

   Synthetic Approach towards the „bent and battered" [2.2]Paracyclophane as a TADF backbone

3) TADF Symposium 09/2017 in Frankfurt/Main

   Towards the "bent and battered" [2.2]Paracyclophane as a TADF backbone

4) European Symposium of Organic Chemistry 07/2017 in Cologne

   [2.2]Paracyclophane-based Functional Surface Coatings by CVD Polymerization

8.3 Acknowledgements

At first, I express my deepest gratitude to my advisor Prof. Dr. Stefan Bräse for his guidance and patience and for giving me this challenging topic. I would like to thank him for his guidance as well as the support and freedom to work and most importantly the numerous opportunities that I was given by him to contribute to the scientific community and to experience what an academic life is like.

Without collaborators, this work would have been much less exciting, therefore I am very grateful to Dr. Eli Zysman-Colman and Nidhi Sharma for the investigation of all luminescent emitters I have provided and I am furthermore sincerely grateful that I was excellently hosted during my research visit in St. Andrews in Oct.-Nov. 2018.

Furthermore, I want to thank Prof. Dr. Joerg Lahann and his group namely Fabian Becker, Dr. Kenneth Cheng and Dr. Christoph Hussal for performing the CVD investigation and collaborating in order to propel our SFB1176 project.

A particular thank you to Dr. Frank Biedermann and Dr. Stephan Sinn who came up with the [2.2]paracyclophane-based host for supramolecular chemist, which – I am sure – will turn out to be an excellent start of an exciting research!

I want to acknowledge the work of Dr. Olaf Fuhr for crystal structure analyses at KIT, Dr. Martin Nieger for crystal structure analyses at the University of Helsinki, Dr. Andreas Rapp and Tanja Ohmer-Scherrer for NMR spectroscopy, Dr. Norbert Foitzik, Lara Hirsch and Angelika Mösle for mass spectrometry and IR. All of them helped me very kindly by analyzing all compounds and without whose work this thesis would not have been possible.

My gratitude also goes to my KSOP mentor Dr. Martin Lauer, who regularly took the time to listen to my problems, helped structuring my time plans and most importantly kept me motivated.

Prof. Dr. Uli Lemmer is kindly acknowledged for the acceptance of the co-reference of this thesis.

To the three graduate schools, KSOP, SFB1176 and 3DMM2O that I was part of, I am sincerely grateful not only for financial support in the form of scholarships, positions of travel funding, but also with regard to the opportunities I was given through networking and personal training events.

I also want to especially thank Jasmin Busch and Christoph Zippel for taking the time to proof read and give me very helpful remarks and suggestions. I know that a doctorate time is stuffed with more things to do than is possible and therefore I am particularly grateful that you took the time to thoroughly correct even under time pressure.

Furthermore, I'd like to thank my Bachelor students Cornelia Mattern and Albert von Orlikowski, who both managed to overcome major synthetic obstacles and therefore helped me a lot with my projects.

A sincere thank you to the "*Hiwi*" students Sarah Al-Muthafer and Moritz Hommrich and the apprentices Lisa Gramespacher and Felix Mazza I was allowed to supervise. Much of the results would not have been possible in this shot time without your help, nor would the time have been that much fun.

I would like to thank the whole AK Bräse team for this excellent time. Thank you, guys, it was my pleasure to be part of the team. In particular, my thanks go to Dr. Stephan Münch and Tobias Bantle for keeping the HPLC alive and to Florian Mohr, Mareen Stahlberger and Zhen Zhang to keep me entertained whenever I've had to spend time in the old building. A huge thanks to Dr. Christin Bednarek, Janine Bolz, Christiane Lampert and Selin Samur, whose help cannot be overestimated in keeping the group running.

A particular thank you goes to the numerous lab mates from different working groups I've had throughout my time in the MZE building, namely Philipp Bohn, Dr. Patrick Dannecker, Dr. Mathias Lang, Yannick Matt, Dr. Dominique Moock, Dr. Christina Retich, Dr. Rebekka Schneider, Rieke Schulte, Kevin Waibel and Dr. Isabelle Wessely and the floor mates, in particular Benjamin Bitterer, Eren Esen, Dr. Zahid Hassan, Roman Nickisch and Julian Windbiel for not only keeping the working days fun and entertaining, but also for valuable advice and help when problems in the daily work of a synthetic chemist arose.

I am very happy to have been part of the "*chicco di caffè*" group with Dr. Claudia Bizzarri, Fabian Hundemer, Dr. Johannes Karcher, Susanne Kirchner, Yannick Matt, and Dr. Zbigniew Pianowski who made every lunch break an outstanding event of the day.

A big thanks to the "Mafia Fanclub" consisting of Dr. Dominik Köhler, Dr. Mathias Lang, and Dr. Dominik Pfaff, with who I got to enjoy the good part of a doctorate time.

I am sincerely grateful to Janina Tolks for providing me with unfailing support and encouragement throughout my years of study.

Finally, I am immensely grateful to my parents for their continuous motivation and neverending support throughout my studies.

Thank you.